GEOMETRICAL FRUSTRATION

This book shows how the concept of geometrical frustration can be used to elucidate the structure and properties of non-periodic materials such as metallic glasses, quasicrystals, amorphous semiconductors and complex liquid crystals.

Geometrical frustration is introduced through examples and idealized models, leading to a consideration of how the concept can be used to identify ordered and defective regions in real materials. Then it is shown how these principles can also be used to model physical properties of materials, in particular specific volume, melting, the structure factor and the glass transition. Final chapters consider geometrical frustration in periodic structures with large cells and quasiperiodic order. Appendixes give all necessary background on geometry, symmetry and tilings. The text considers geometrical frustration at different scales in many types of materials and structures, including metals, amorphous solids, liquid crystals, amphiphiles, cholesteric systems, polymers, phospholipid membranes, atomic clusters and quasicrystals.

This book will be of interest to researchers in condensed matter physics, materials science and structural chemistry, as well as mathematics and structural biology.

JEAN-FRANÇOIS SADOC is Professor at Paris-Sud University where he teaches crystallography and statistical physics. At the 'Laboratoire de Physique des Solides' in Orsay, his main fields of interest are disordered systems. In particular, he focuses on liquid crystal structures and metallic quasicrystals. In 1983 he was awarded the Winter–Klein prize for his work on metallic glasses by the French Academy of Sciences.

RÉMY MOSSERI is Director of Research at the 'Centre National de la Recherche Scientifique' and studies the structural and electronic properties of quasicrystals at the 'Groupe de Physique des Solides' at Paris University in Jussieu. He is also interested in the history of physics. In 1992 he was awarded the Paul Langevin prize by the French Physical Society.

Geometrical Frustration

JEAN-FRANÇOIS SADOC

Université de Paris-Sud

RÉMY MOSSERI

Université Paris – Jussieu

CAMBRIDGE
UNIVERSITY PRESS

CAMBRIDGE UNIVERSITY PRESS
Cambridge, New York, Melbourne, Madrid, Cape Town, Singapore, São Paulo

Cambridge University Press
The Edinburgh Building, Cambridge CB2 2RU, UK

Published in the United States of America by Cambridge University Press, New York

www.cambridge.org
Information on this title: www.cambridge.org/9780521441988

First published 1999
This digitally printed first paperback version 2006

A catalogue record for this publication is available from the British Library

Library of Congress Cataloguing in Publication data
Sadoc, J. F.
    Geometrical frustration / Jean-François Sadoc and Rémy Mosseri.
        p.   cm. – (Collection Aléa-Saclay. Monographs and texts in
    statistical physics)
        ISBN 0 521 44198 6 (hc.)
        1. Statistical physics.   2. Condensed matter.   3. Materials.
    I. Mosseri, Rémy.   II. Title.   III. Series.
    QC174.8.S25   1999
    530.13–dc21   98-39488 CIP

                    ISBN-13 978-0-521-44198-8 hardback
                    ISBN-10 0-521-44198-6 hardback

                    ISBN-13 978-0-521-03187-5 paperback
                    ISBN-10 0-521-03187-7 paperback

Contents

Preface

Geometrical Frustration! Such a title requires at least two preliminary remarks. First, this book is about condensed matter physics, and in particular cases where matter is spatially organized in complex structures like large cell crystals, glasses, quasicrystals, and also some liquid crystal organizations. The second remark concerns using the word 'frustration'. Let us immediately rule out its psychological meaning, and keep only its technical one, introduced about 20 years ago in the context of spin glasses, and which subsequently diffused to neighbouring fields, thanks to physical analogies. Indeed, we hope to demonstrate that, studying 'frustrated systems', far from generating any frus- tration, will create satisfaction linked to a better understanding of a rich and complex domain.

Geometrical frustration covers situations where a certain type of local order, favoured by physical interactions, cannot propagate throughout space. A classical example is that of pentagonal, or icosahedral, order which appears in the three-dimensional sphere packing problem. This symmetry, which is not compatible with translations, is nevertheless met, often imperfectly, in numerous materials. Such strong contradictions between local and global configurations are found in various physical systems, with different kinds of interactions, and coherence sizes. The concept of frustration then applies to metallic alloys at the atomic scale, to liquid crystal organizations like amphiphiles films or cholesteric blue phases at the scale of hundreds of thousands of ångströms, and even to larger scales with some biological structures.

The unifying character of this concept of geometrical frustration appears also in the very general method used to describe these complex systems at a theoretical level. It uses curved space, essentially a hypersphere with adequate radius, where local and global order become compatible. This book presents in some detail these different ideal models, as well as the main geometrical tools for manipulating them.

Condensed matter physics deals largely with the relation between structure and properties. It is well established now that the latter are sensitive to the ordered structure itself as well as to the defects. The approach developed here allows for a better definition of defects in non-periodic systems. It distinguishes between what will be called intrinsic defects, which are unavoidable and linked to the presence of frustration, and more extrinsic defects, closer to the type of defects encountered in simple crystals. Defects will be characterized upon analysing the way of recovering Euclidean space starting from the ideal structures in curved space. Among these defects, disclination lines play a major role.

We have chosen mainly to present here our own contribution to this subject, sometimes in a rather detailed form, and to summarize briefly, or sometime only quote, the other works. This is the occasion to thank all those with whom we have been collaborating, or simply with whom we had some illuminating discussions, knowing that, dealing with a period of about fifteen years, we will certainly forget to cite some of them. We therefore thank Francis Bailly, Yves Bouligand, Marc Brodsky, Jean Charvolin, Marianne Clerc, Rossen Dandoloff, Nicolas Destainville, David DiVincenzo, Jean Dixmier, Elisabeth Dubois-Violette, Michel Duneau, Jacques Friedel, Françoise Gaill, Jean-Pierre Gaspard, André Guinier, Rémi Jullien, Maurice Kléman, Alan Mackay, David Nelson, Stam Nicolis, Christophe Oguey, Brigitte Pansu, Nicolas Rivier, Anne Sadoc, Clément Sire and Mike Widom.

Claude Godrèche was present at the inception of this book; we thank him for his enthusiasm and efficiency.

We finally would deeply like to thank our families, who were patient and helpful during all these years.

Jean-François Sadoc and Rémy Mosseri
February 1999

1

Introduction to geometrical frustration

1.1 From cubism to icosahedrism

Among the many scientific breakthroughs which have brought solid state physics to be considered as a distinct field of physics, with its own concepts and methods, the most important one was probably the discovery, by Max Von Laue, Walter Friedrich and Paul Knipping in 1912, of the diffraction of X-rays by a crystal. It proved that crystals were made of a periodic array of atoms or molecules, an idea which was already supported by the work of abbé René-Juste Haüy at the end of the eighteenth century. Indeed, the latter proposed a periodic microscopic structure for crystals, based on the observation of the regular facets of crystal grains at a macroscopic level.

The mathematicians of the nineteenth century contributed to this story by inventing a very important tool, the concept of a transformation group, which would prove useful in almost every field of physics. As far as the groups of space symmetry operations are concerned, the complete classification of the spatial groups was fulfilled by the end of the nineteenth century for the three-dimensional Euclidean space. A very important result which follows this classification is the crucial restriction on the compatibility between rotations and translations: only rotations of order 2, 3, 4, 6 can let a crystal be invariant. An important ingredient in the description of an ordered structure is its point group, which enumerates the symmetry operations which leave a point fixed. The natural point groups in the three-dimensional Euclidean space are those associated with the five regular Platonic solids, the tetrahedron, octahedron, cube, icosahedron and dodecahedron (figure 1.1). But the latter two share the same symmetry group, of order 120, which contains five-fold rotations and therefore cannot accommodate translations. This is why crystallographers ignored this symmetry for almost half a century. It is fair to say that they had enough work with the analysis of thousands of crystalline structures, the knowledge of atomic positions being a prerequisite for most attempts in

1

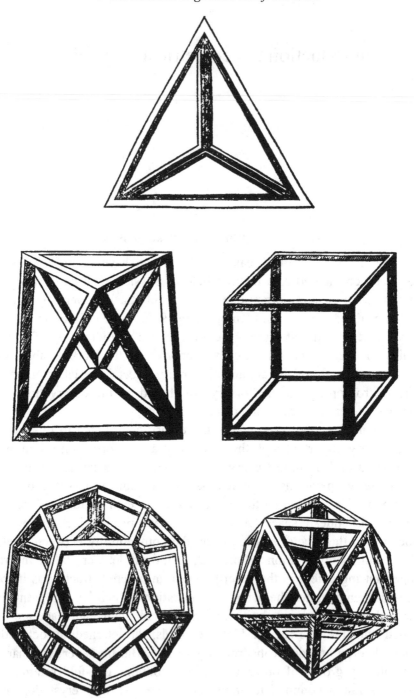

Fig. 1.1. The five Platonic solids: tetrahedron, octahedron, cube, dodecahedron and icosahedron.

explaining the physical properties. For instance, carbon is found with two crystalline forms, graphite and diamond, which have very different physical properties resulting from different local atomic organization. It is fascinating to see that many (if not all) of the 230 crystal groups have a concrete realization in Nature. We shall see later in this book that some 'forbidden' symmetries fall also into that class, at the price of a subtle redefinition of what 'order' means.

Contrary to the history of experimental crystallography, the rise of interest in icosahedral symmetry is more diluted in time, and therefore difficult to date. At least we can say that this interest is much older than the discovery of quasicrystals, although the latter is a turning point for the role played by this symmetry in physics. Almost absent in the first half of the twentieth century, it is still often presented very briefly (if not ignored) in most textbooks (previous to 1980). Note however the discussion of five-fold and icosahedral symmetry in Kittel's *Introduction to Solid State Physics* (1966).

But, icosahedral symmetry appears as a point group for molecules, like the now famous C_{60} fullerenes, and in small size clusters of metals and rare gases. It also manifests itself as a local arrangement in some metals and alloys in their liquid and amorphous structures, and even in some large unit cell crystals, the Frank–Kasper phases (1958), as part of the repetitive crystalline motif. Some covalent structures, like boron, or hydrogen bonded clathatres, also present local icosahedral atomic configurations.

Among others, Linus Pauling (1960, 1985) contributed considerably to the understanding of such structures. It is from this point of view quite interesting to remark that he was not very enthusiastic in considering quasicrystals as a new type of structure and tried to make them fit older classification schemes. The prestigious English school of metallurgy, which gave important contributions to the understanding of atomic and electronic structures in crystals (Hume-Rothery (1926), Jones (1960), Mott, Frank, Kasper and so many more), was also influential in the rising interest in icosahedral symmetry, like J.D. Bernal (1964), who was interested in modelling of liquid structures. In France it is mainly through work concerning amorphous metals (Sadoc *et al.* 1973) and rare gas aggregates (Farge *et al.* 1975) that five-fold symmetry was popular in the solid state community.

Outside the fields of physics and chemistry, it is clearly in biology where icosahedral symmetry occurs, for instance in the geometrical description of many viruses. It is probably for the functional necessity to be compact and to separate an inside and an outside, that biological structures adopt this symmetry, which approximates the sphere better than any other. But icosahedral symmetry is not limited to the scientific world. One can even think that it has been mainly popularized by the Dutch artist Escher, who used it in many drawings, and the American engineer and designer Buckminster Fuller, through his famous geodesic domes.

More recently, icosahedral symmetry came on the scene in at least two spectacular manifestations. In 1985, came the discovery of a new ordered structure for carbon (see figure 1.2), made of a periodic packing of large molecular units of C_{60} (Kroto *et al.* 1985). Notice that in reference to Buckminster Fuller this new molecule is often called 'fullerene'. This new domain seems rich for potential applications.

But, even more important at the fundamental level, was the discovery in 1984 of an icosahedral quasicrystal, in AlMn metallic binary alloys, which is a new class of non-crystalline, but nevertheless ordered, structures in solids (Shechtman *et al.* 1984).

The starting point of crystallography, in 1912, and the culminant event in icosahedrism, in 1984, both marked by a crucial experiment, show an interesting difference at the historical level.

The diffraction of X-rays by crystals, which had such a big impact in physics, was probably the most important experimental discovery ever made in a theoretical institution. Indeed, although the X-ray tube was borrowed from the Institute of Experimental Physics of Munich, chaired by Röntgen, the

Fig. 1.2. Drawing of a C_{60} molecule: notice the local six-fold but also five-fold arrangements of atoms.

experiment was performed at the neighbouring Institute of Theoretical Physics whose director was Arnold Sommerfeld. The intention was of fundamental nature, to prove the undulatory nature of the X-ray radiation; having, in addition, created a new and important field of research, Laue and his collaborators therefore killed two birds with one stone.

The discovery of quasicrystals, 72 years later, was to some extent, the answer from the experimentalists. Indeed, for a limited period, it seemed to disprove a belief which had the status of a theorem among most crystallographers; that order, in the form of a point-like diffraction pattern, is incompatible with five-fold symmetry (see figure 1.3). It was then quickly recognized that the solution to this apparent contradiction was to be found in the field of quasiperiodic tilings similar to those introduced in 1974 by Roger Penrose (see Gardner 1977).

So, icosahedral order is the first example we meet where a kind of local order is not compatible with regular space filling. But the theme of the present book, *Geometrical Frustration*, covers a much wider range of systems, with different types of order, which all share this initial incompatibility, called frustration; but they can also be treated along an unified approach, with the help of so-called ideal models, most often defined in a three-dimensional space of positive curvature, the sphere S^3.

The theory described here is not only relevant for disordered materials, but also for several classes of crystalline structures like Frank–Kasper phases (Nelson 1983, Sadoc 1983), clathrates, and even complex fluids like cholesteric blue phases (Sethna *et al.* 1983) and amphiphilic membranes (Sadoc and Charvolin 1986). The concept of geometrical frustration is expected also to play a role in biological systems, either at the scale of biological tissues (Bouligand 1990) or at the scale of molecules (Bascle *et al.* 1993).

The present approach is mainly concerned with the structural aspects, even if their influence on physical properties is also considered, as in chapter 6. The deep interplay between physics and mathematics has often been emphasized, and the natural tool for treating atomic structures is geometry. The latter plays an important role in this book, at a level of complexity which is probably higher than the usual level encountered in solid state physics. This explains why, even in this introductory chapter, it is given a prominent place.

1.2 Geometry

1.2.1 The underlying space

The fact that a physical system should obey a principle of minimum energy is always constrained by external conditions (temperature, pressure, geometrical

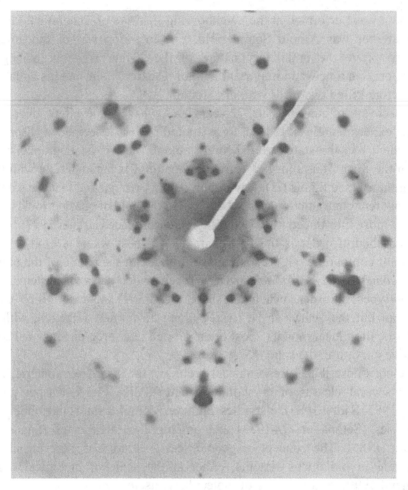

Fig. 1.3. Diffraction pattern for an AlLiCu quasicrystal of icosahedral symmetry oriented along a five-fold axis (courtesy of F. Denoyer).

confinement, ...). One such condition, which is so evident that it is often forgotten, is the type of space underlying the system. The space is characterized by several quantities of topological and metrical nature, which we are going to describe now. The importance of such knowledge comes from the fact that, in some cases, the local atomic configuration which minimizes a local form of the energy, may not be compatible with the underlying space (it is not 'space-filling').

1.2.2 The dimension

There are several definitions for the dimension, which has been thoroughly discussed since the enormous growth of interest in fractal spaces, whose dimension takes non-integral values. The simplest definition is maybe to continuously distribute mass in the space and to see how the total mass scales with the size. But there are many problems connected to this question of dimension which we shall not discuss here.

Here we shall be interested in spaces of integral dimension, called manifolds. By definition, a d-dimensional manifold is a space which can be locally represented by a d-dimensional Euclidean space. Note that we shall not always give here the precise mathematical definitions for the different terms. The technical definitions, which are undoubtedly essential for precise studies, may obscure this brief presentation especially for those unfamiliar with basic topological objects and tools. The interested reader can go further on this subject with the celebrated book *Geometry and the Imagination* by Hilbert and Cohn-Vossen (1952).

1.2.3 Topological description

In a topological approach, two objects which can be continuously deformed into one another will not be distinguished. For instance a sphere and an ellipsoid are the same object while a torus belongs to a different class. Note that a sphere denotes the 'surface' defined by

$$x_1^2 + x_2^2 + x_3^2 = R^2$$

in Euclidean three-dimensional space, while its interior is called a 'ball'. Hence the usual sphere is a two-dimensional manifold, denoted S^2.

At first sight the topological classification might appear crude (even if topology is a rather recent field of mathematics) but in fact it already displays deep properties. For instance it was proved in the mid-nineteenth century that every compact orientable two-dimensional manifold is topologically equivalent to the surface of a 'doughnut' with some number of holes. This number is called the genus of the surface.

For example, the sphere S^2 is a surface of genus 0, the usual torus has genus 1, etc .... The genus is related to the maximum number of (topologically) distinct closed curves on that surface which do not separate it into two disjoint parts. When the surface is tiled with polygonal faces of arbitrary shape, there is an invariant quantity relating the number of vertices (V), edges (E) and faces (F) which depends only on the surface:

$$V - E + F = 2 - 2g = \chi$$

where g is the genus and χ is called the Euler–Poincaré characteristic of the surface (Coxeter 1969, Hilbert and Cohn-Vossen 1952). This relation is very powerful and leads to important and simple results on the average properties of two-dimensional tilings as we shall see later. It has been extended to higher dimensional polyhedral subdivision of the *n*-dimensional sphere S^n in the form:

$$\sum (-1)^i A_i = 1 + (-1)^n$$

where A_i is the number of *i*-dimensional 'cells' in the tiling (Poincaré 1893).

1.2.4 Geometrical description

In the history of geometry one can recognize two main approaches, the local and the global, which culminate in the nineteenth century with Klein (1900) and Riemann (1892).

Geometry, in Klein's view, is the study of the properties of figures which are conserved when the whole space is subject to a group of transformations, e.g. to build the theory of invariants with respect to this group. This leads to an 'order' among the groups and their associated geometry. At the top is the projective geometry with projective transformations keeping invariant the collinearity of points, the harmonic ratio of points, the degree of algebraic curves .... Then comes the affine geometry (the affine group being a subgroup of the projective group) which, in addition to the previous properties, conserves parallelism. Note that affine transformations are projective transformations which set as invariant the 'line at infinity' of the projective geometry. In the same way the metric geometries (in two dimensions) correspond to a subgroup of the projective group which set as invariant a conic of the projective plane. According to whether this conic, called the absolute, is real, imaginary or degenerate (reduced to two imaginary points), the geometry will be hyperbolic, elliptic or Euclidean. The classification of Klein is not limited to two or three dimensions but applies to manifolds of arbitrary dimension. Note that whatever the invariant properties are, the main characteristic of Klein spaces is their homogeneity. The group acts indistinguishably in any point of the manifold.

The spaces of Riemann are, on the contrary, devoid of any kind of homogeneity. They are characterized at each point of the manifold by a quadratic form:

$$ds^2 = \sum g_{ij}\, du_i\, du_j$$

which defines the square of the distance between two infinitesimally close neighbouring points and generalizes the Euclidean formula:

$$ds^2 = du_1^2 + du_2^2 + \cdots + du_n^2$$

where g_{ij}, the metric tensor, is *a priori* arbitrary and varies from point to point. As a result, in a Riemann space, two observers can locate points in their immediate neighbourhood but cannot, without new conventions, relate a coordinate frame with respect to another. The generic Riemann space is then a juxtaposition of unrelated pieces. It is possible to give more structure to a Riemann space by introducing a 'connection' which allows the comparison of different reference frames by translating them along paths on the manifold (see appendix A8, §3.2).

The distinction between Klein and Riemann geometry is an example among others of a constant duality of point of view occurring in mathematics (Lautmann 1977) and more generally in science: the opposition between local and global approaches. Such a distinction is also present in physics between the restricted and the generalized relativity theory. The former is concerned with the four-dimensional Minkowski space by the study of invariants of the Lorentz group, while the latter, of Riemann type, relates the local value of g_{ij} to the distribution of matter at this point, the invariance of physical laws being encoded in the 'connection' of that space.

At a different level we shall face similar questions in our study of atomic structures. Regular structures are such that no contradiction exists between local and global requirements, in which case the global approach (with symmetry groups) is shown to be very powerful. The kind of regularity in a crystal is the discrete version of the homogeneity of a Klein space. In less regular structures, the local configuration may be viewed in some cases as the discrete analogue of a quantity which can be calculated from the g_{ij}, that is to say the local curvature. Defining an ideal structure where the local configuration can propagate is then equivalent to finding a new geometry with the appropriate distribution of curvature. If such geometry allows for a global description (Klein geometry), this ideal model is again regular and can be studied on its own. The relation between the initial structure and the ideal one is studied under different types of mapping. This point of view is called the 'curved space model' of disordered systems and will be discussed at length here. Note that we have passed very quickly over the question of analogy between continuous and discrete problems. In two dimensions this is rather simple, thanks to the existence of the Gauss–Bonnet relation (chapter 4, §3.1). This is much more complex in three dimensions where approximations have to be done in order to get quantitative results (chapter 4, §3.4).

1.3 Geometrical frustration

1.3.1 From local order to global space filling

A main question that a condensed matter physicist faces is to explain the stability of a solid. Already in molecules, the most precise quantum mechanical calculations often show large diversity for the low energy atomic geometries. Because of their size, solids require that many approximations be done in order to compute their cohesive energy.

Even if, in recent years, some progress has been made in combining quantum mechanics and numerical simulations in order to investigate the configuration space of solids, we still need to approximate this energy as a sum of finite-range interactions, like two- or many-body potentials. This is not always possible, some subtle structural effects being explained only on the basis of quantum mechanical arguments. Nevertheless, it is often possible to establish some local rules, of a chemical nature, which lead to low energy configurations and therefore govern structural and chemical order.

It is then crucial to analyse to what extent a low energy local configuration can be assigned to each atom or molecule; clearly, if it is possible, according to the additive property for the energy in this type of classical model, this should give at once the ground state of the solid. The subject of this book refers to the opposite case, called 'geometrical frustration', when the local order cannot be propagated freely throughout the space. The simplest example, which will be detailed at length below, is the case where the local rule consists of packing spheres as densely as possible, a crude model for metallic atoms.

In two dimensions, a local packing of discs, the flat analogues of spheres, is densely organized if centres of discs are located on vertices of equilateral triangles, which can tile the plane along the six-fold symmetric triangular lattice; this is an unfrustrated case. In three dimensions, the local densest packing of spheres is achieved by placing their centres at vertices of a regular tetrahedron. The geometrical frustration is revealed immediately in that the three-dimensional Euclidean space cannot be filled completely by regular tetrahedra (see chapter 2).

Our definition of geometrical frustration is general enough not to be restricted to the atomic level, or even to discrete systems. For example, in liquid crystals, under a continuous approximation, it is possible to characterize some systems as being geometrically frustrated: phases of cholesteric twisted molecules or assemblies of amphiphilic bilayers are two cases which will be presented below.

A common feature of all these systems is that, even with simple local rules, they present a large set of, often complex, structural realizations. This recalls another area of physics, that of frustrated spin systems (Toulouse 1977), whose complex energy landscape has been the subject of many deep contributions in the past ten years. The main difference with the purpose of this book is that it is not the location in space of the spins which is subject to variation, but the value of the internal variable, the spin itself. As a consequence, even with a common problem, the method of theoretical investigation largely differs.

Here, we shall focus mainly on geometrical tools. We hope to convince the reader that geometrical frustration is a *unifying concept*, which plays a crucial role in very different fields of condensed matter, ranging from clusters and amorphous solids to complex fluids. In all cases, we propose a method of approach which follows two steps. First, the constraint of perfect space-filling is relaxed by allowing for space curvature. An ideal, unfrustrated, structure is defined in this curved space. Then, specific distortions are applied to this ideal template in order to embed it in the three-dimensional Euclidean space. It allows us to describe the final structure as a mixture of ordered regions, where the local order is similar to that of the template, with defects arising from the embedding. Among the possible defects, disclinations will play an important role.

1.3.2 Complexity out of simple rules

Having found 'hidden order' in structurally disordered matter, one can ask whether this is a unique situation, or if similar patterns are present in other fields. Indeed, recent years have shown an increasing interest for systems with a high degree of complexity in different branches of physics. Complexity is not always due to the presence of many independent parameters but can appear in *a priori* simple cases. The paradigmatic example is the dynamical system associated with the logistic map (iteration of $x_{n+1} = ax_n(1 - x_n)$), which displays a variety of regimes, from periodic to chaotic, as the parameter a is varied from zero to four. A second example is the 'game of life', invented by J. Conway, where two-dimensional cellular structures evolve under the same very simple proximity induced rearrangements into very different patterns.

But such complexity is not limited to purely theoretical examples. Coming to our subject, a student in crystallography could already be quite puzzled to learn that, among possible structures, some are relatively simple, like the f.c.c. structure for dense metals, but that, even for pure monoatomic metals, rather intricate structures of a high complexity can occur. As an example, manganese crystallizes in a structure with a unit cell containing 58 atoms! Here, the local rules, if the problem can be simplified to local interactions, should explain such a structure. And indeed, as will be shown below, our approach, originally developed to describe amorphous materials, can also shed new light on these types of complex crystalline structures.

In several complex systems, a better understanding has been reached by introducing a new point of view and/or new variables. This role has been played by adding, to the three-dimensional Euclidean underlying geometry, extra curvature (amorphous systems) or dimensions (quasicrystals). New bridges appear between otherwise independent subjects. An interesting example is the analogy between time dependence in certain dynamical systems and spatial order in solids. The former presents systems with periodic, quasiperiodic and chaotic behaviours. The first two are clearly the time versions of crystals and quasicrystals and the last one has been investigated as the paradigm for amorphicity, in terms of spatial chaos (Reichert and Schilling 1984).

For years, probability theory was almost the only mathematical branch involved in the study of disordered systems. The situation has now changed and other fields of mathematics are now used, about which solid state physicists are still often unfamiliar: fractal geometry, number theory, topology, other particular geometries like curved spaces, higher dimensional spaces, ultrametricity ...). As so often in mathematics these different branches are interconnected and might be the source of the above mentioned analogies between different physical systems.

1.3.3 Structural complexity encoded into defect networks

Anticipating the content of this book, we can announce one of its important qualitative results on complex frustrated structures: it is possible to save one order of magnitude for their description at the microscopic level. Since it is possible to define an 'order' in these frustrated systems, it becomes possible to define defects, whose characteristic scale depends on the degree of frustration.

The complexity of the structure is then encoded into the complexity of its defect network (which are mainly disclination lines). One then meets:

- ideal systems, without defects and frustration, which are described in chapter 2;
- structures with a periodic network of defects, such as the large cell crystals described in chapter 7;
- structures with hierarchical networks of defects, such as the polytopes described in chapter 5 or the quasicrystals of chapter 8;
- and finally amorphous materials, characterized by a disordered network of defects.

1.3.4 A diverse community

The concept of geometrical frustration has emerged from the work of a whole community of solid state physicists working in the field of disordered structures, but their reflections were enriched by looking to other fields: in physics, through theoretical tools developed in the context of general relativity, in mathematics, with many geometrical concepts, but also in chemistry, biology ....

Even if this book is devoted, to a large part, to a presentation of the essential features of our own work on geometrical frustration, the ideas which are developed here result from confrontation with and contributions from this rather diverse community.

In condensed matter physics, the concept of topological defects is now well known as extremely fruitful. Modern ideas, some of which are rooted in metallurgy, have also emerged, or been used, in this rather different field of physical chemistry, often called 'soft matter', which covers liquid crystals, polymers .... Maurice Kléman has strongly contributed to establishing this link. A place like the 'Laboratoire de Physique des Solides d'Orsay', where metallurgists and physico-chemists are gathered, if not a condition, was certainly helpful in establishing this unifying concept of geometrical frustration (see, for instance, Dubois-Violette and Pansu 1988, Charvolin and Sadoc 1994).

Researchers from the statistical physics community also contributed in the same direction, often at the price of manipulating complex atomic structures which are usually more familiar to crystallographers. Nicolas Rivier, who developed a gauge theory of disordered systems, was also the first to introduce the idea of topological disclinations in glasses (Rivier 1979, 1983). Another fruitful endeavour is represented by David Nelson and his collaborators (Nelson and Widom 1984), with an initial interest in the theory of three-dimensional melting.

One should quote here a very pedagogical and interesting book, *Beyond the Crystal* by Venkataraman *et al.* (1989), which is devoted to questions closely related to geometrical frustration.

The present book follows a track which insists on the unifying aspect of the theory: a description of the different kinds of curved space ideal models, followed by a discussion on mapping and defects, and finally a summary of some physical consequences. But some readers could gain more profit in using it less sequentially, focusing on one kind of material instead of jumping for instance from metals to liquid crystals. In addition, for ease of reading, we have put the presentation of most of the mathematical tools into several appendixes.

It is by going back and forth between these different fields, and by discussion with specialists who do not share necessarily the same language, that these ideas have matured, and are still progressing. But, isn't it the natural way for scientific reflection to proceed?

2

Ideal models

2.1 A unified approach to very different materials

It is not *a priori* surprising that an approach whose nature is mainly geometrical can describe very different materials, like metallic and covalent solids, amphiphilic molecules, cholesteric blue phases, etc .... As far as crystals are concerned, 'mathematical' crystallography takes this role of classifying all the possible structures compatible with periodicity in space, without regard to the chemical nature of the atoms. The physicist enters the field in order to relate an experimental pattern, usually the result of a diffraction experiment, to one of the possible periodic arrangements. This is already an idealization in the sense that most of the studied materials are only imperfect crystals, containing many types of defect. A perfect crystal is impossible for the first reason that it is necessarily limited in size. As is well known now, atoms close to the surface rearrange for the sake of energy minimization. A second reason is temperature: when T is finite, vacancies can contribute to configurational entropy. But standard, bulk diffraction is usually blind to these alterations of the periodic order. A natural and unavoidable obstacle to a precise structural determination is the experimental precision. The wave probes (X-rays, electrons, neutrons) have limited wave vectors, which introduce cut-like effects in the passage from reciprocal to real space.

This may not be important for simple crystals, but can prevent a precise determination for crystals with large unit cells. The precision of the wave detector is also crucial, in that the natural peak width is directly connected to the range of positional order and must therefore be distinguished from the apparatus width resolution. The physicist can take this into account by defining what he or she means by a 'physical' perfect (mono)crystal (as opposed to a mathematical one). It is a very large, but nevertheless limited, periodic arrangement of atoms, without internal grain boundaries, such that its diffraction pattern can be indexed by a space group, with peak widths smaller than the apparatus resolution. Note that this neither imposes that it is defect-free and nor that, for large cell crystals, the precise atomic position be known. These questions of definition were recently in the forefront with the discovery of 'perfect' quasicrystals, whose diffraction shows resolution limited peak widths. Note that in order to include liquid crystals in this description, we should again lower our definition, and simply require that the two-point density correlation function be periodic.

Equipped with the definition of a perfect crystal, either mathematical or physical, it is then possible to address the question of defects, defined with respect to this perfect crystalline order. The study of defects in solids has been one of the favourite subjects in solid state physics during the second half of the twentieth century, once it was fully appreciated that they often play a prominent role in physical properties. One is then naturally brought to ask questions about the nature and role of defects in non-crystalline solids. But before being able to recognize defects in non-crystalline materials, we must give a precise definition of the kind of order to which these defects refer. It is the aim of this chapter to present the 'ideal' models that can be associated with the frustrated systems. This will take us into the fascinating world of non-Euclidean geometries (see also appendix A1), especially positively curved space, the hypersphere S^3. But to be better prepared for such an abstraction, we shall first discuss simpler examples.

2.2 Simple two-dimensional examples

Two simple two-dimensional examples are helpful in order to get some understanding about the origin of the competition between local rules and geometry in the large scale. Consider first an arrangement of identical discs (a model for a 'hypothetical' two-dimensional metal) on a plane; we suppose that the interaction between discs is isotropic and locally tends to arrange the discs in the densest way possible. The best arrangement for three discs is trivially an equilateral triangle with the disc centres located at the triangle vertices. The study of the long range structure can therefore be reduced to that of plane tilings with equilateral triangles. A well known solution is provided by the triangular tiling: we find a total compatibility between the local and global rules and the system is said to be 'unfrustrated'.

In our second two-dimensional example, the interaction energy is supposed to be at a minimum when atoms sit on the vertices of a regular pentagon. Trying to propagate in the long range a packing of these pentagons sharing edges (atomic bonds) and vertices (atoms), one faces a difficulty due to the impossibility of tiling a plane with regular pentagons, simply because the pentagon vertex angle does not divide 2π as an integer.

Three such pentagons can easily fit at a common vertex, but a gap remains between two edges. It is this kind of discrepancy which is called 'geometrical frustration'. Note that the term 'frustration' is used to emphasize that a goal cannot be reached (to perfectly tile the plane), but not with respect to the large and fascinating field of investigation that is opened, as we hope to convince the reader!

A long time ago, Kepler (1619) played with this pentagon tiling problem, and introduced other five-fold symmetric figures to fit into the holes between the pentagons, providing very nice anticipations of the Penrose tilings; see figure 2.1.

There is one way to overcome this difficulty. Let the surface to be tiled be free of any presupposed topology and metrics, and let us build the tiling with a

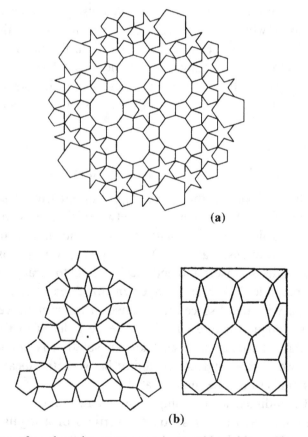

(a)

(b)

Fig. 2.1. Tiling of a plane by pentagons is an old problem. Shown here are some solutions proposed by Kepler (a) and Dürer (b).

strict application of the local interaction rule. In this simple example, we observe that the surface inherits the topology of a sphere and so receives a curvature (figure 2.2). The final structure, here a pentagonal dodecahedron, allows for a perfect propagation of the pentagonal order. It is called an 'ideal' (defect-free) model for the considered structure. This will be our constant strategy: to solve the frustration by allowing the use of new underlying geometries.

Anticipating the following, we can make some remarks on the chosen rule. It consists of propagating a pentagonal order with three pentagons meeting at a vertex. Why three, and not four? The reason is that this choice minimizes the change of the underlying space (the curvature): the angle at a vertex, 108°, is rather close to 120°. If we choose to put four pentagons at a vertex, the underlying space inherits negative curvature and becomes a hyperbolic plane. An important difference between these two possible ideal models is that the model on the sphere contains a finite number of tiles, while this number is infinite for the hyperbolic plane.

2.3 Metals

2.3.1 Dense structures and tetrahedral packings

The stability of metals is a long-standing question of solid state physics, which can only be understood in the quantum mechanical framework by properly taking into account the interaction between the positively charged ions and the valence and conduction electrons. It is nevertheless possible to use a very simplified picture of metallic bonding and only keep an isotropic type of

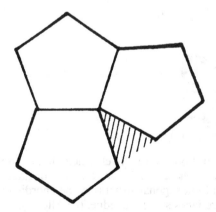
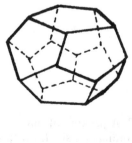

Fig. 2.2. Tiling of a plane by pentagons is impossible but can be realized on a sphere in the form of a pentagonal dodecahedron.

interaction, leading to structures which can be represented as densely packed spheres. And indeed the crystalline simple metal structures are often either close packed face centred cubic (f.c.c.) or hexagonal close packing (h.c.p.) lattices. Amorphous metals can also be modelled by close packing of spheres. But a main difference is clearly seen from the analysis of X-ray diffraction experiments (Sadoc *et al.* 1973). The local atomic order is well modelled by a close packing of tetrahedra, leading to an imperfect icosahedral order.

A regular tetrahedron is the densest configuration for the packing of four equal spheres. The dense random packing of hard spheres problem can thus be mapped on the tetrahedral packing problem. It is a practical exercise to try to pack table tennis balls in order to form only tetrahedral configurations (figure 2.3). One starts with four balls arranged as a perfect tetrahedron, and tries to add new spheres, while forming new tetrahedra. The next solution, with five balls, is trivially two tetrahedra sharing a common face; note that already with this solution, the f.c.c. structure, which contains individual tetrahedral holes, does not show such a configuration (the tetrahedra share edges, not faces). With six balls, three regular tetrahedra are built, and the cluster is incompatible with all compact crystalline structures (f.c.c. and h.c.p.). Note however that, in terms of number of contacts between spheres, an octahedral configuration would

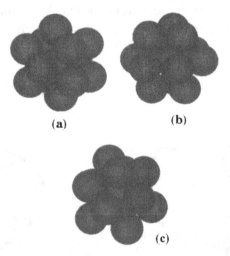

(a)						(b)

(c)

Fig. 2.3. A packing of thirteen spheres; a central one and its twelve neighbours on the first coordination shell. (a) In f.c.c. structures, sphere centres are on cubo-octahedron vertices with triangular and square faces. (b) In hexagonal structures, the coordination shell is close to a cubo-octahedron, but square faces share one edge. (c) Allowing only tetrahedral configurations leads to a distorted icosahedral coordination, all faces being triangular.

compete here, for $N = 6$, with a tetrahedral close packing; but adding more spheres will again favour the latter: adding a seventh sphere gives a new cluster consisting of two 'axial' balls touching each other and five others touching the latter two balls, the outer shape being an almost regular pentagonal bi-pyramid. However, we are now facing a real packing problem, analogous to the one encountered above with the pentagonal tiling in two dimensions. The dihedral angle of a tetrahedron is not commensurable with 2π; consequently, a hole remains between two faces of neighbouring tetrahedra.

As a consequence, a perfect tiling of the Euclidean space R^3 is impossible with regular tetrahedra. Note that, at this local level, the deviation from perfectness is of a metrical nature (figure 2.4). So, in the case of polytetrahedral structures the frustration is due to the impossibility to tile the space with regular tetrahedra. Slightly softening the spheres will not help. Indeed, as we shall see further on, the frustration has also a topological character: it is impossible to fill Euclidean space with tetrahedra, even severely distorted, if we impose that a constant number of tetrahedra (here five) share a common edge.

The next step is crucial: we define an unfrustrated structure by allowing, as in the two-dimensional example, for curvature in the space, in order for the local configurations to propagate identically and without defects throughout the whole space.

2.3.2 Regular packing of tetrahedra: the polytope {3, 3, 5}

Twenty tetrahedra pack with a common vertex in such a way that the twelve outer vertices form an irregular icosahedron (figure 2.4c). Indeed the icosahedron edge length l is slightly longer than the circumsphere radius r

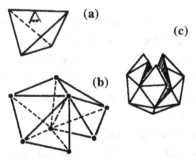

Fig. 2.4. Tetrahedral packing. (a) A tetrahedron. (b) The dihedral angle of a tetrahedron is not commensurable with 2π; consequently, a hole remains between two faces of a packing of five tetrahedra with a common edge. (c) A packing of twenty tetrahedra with a common vertex in such a way that the twelve outer vertices form an irregular icosahedron.

($l \simeq 1.05r$). It is possible to make this icosahedron regular by first shortening accordingly the internal tetrahedral edges. But the price to pay is that the tetrahedra are now no longer regular. In terms of ball packing, this means that the central ball is given a smaller radius than its twelve neighbours. Note that such a configuration is sometimes found, in real metallic alloys.

The tetrahedral regularity can then be recovered by giving a small fourth coordinate to the central site. One then tries to add new shells of sites while keeping the same icosahedral environment for all the sites. It is possible if these new sites are also given at a fourth coordinate with opposite sign. This procedure can be continued for several shells until one sees that no new sites are needed. Indeed a finite set of points has been generated such that each vertex has twelve neighbours in perfect icosahedral configuration. Let us describe this set.

– There are 120 vertices which all belong to the hypersphere S^3 with radius equal to the golden number ($\tau = (1 + \sqrt{5})/2$) if the edges are of unit length.
– The 600 cells are regular tetrahedra grouped by five around a common edge and by twenty around a common vertex.
– This structure is called a polytope (Coxeter 1973a, b) which is the general name in higher dimension in the series polygon, polyhedron, ...
– Even if this structure is embedded in four dimensions, it can be considered as a three-dimensional figure in the following sense. The hypersphere S^3 is defined by the equation

$$x_0^2 + x_1^2 + x_2^2 + x_3^2 = r^2 \tag{2.3.1}$$

which shows that, among the four coordinates, only three are independent. As long as one disregards its interior, the 'hyper-ball' limited by S^3, the latter can be considered as a three-dimensional (curved) manifold. This is analogous to considering the usual sphere S^2 as a curved surface.

This point is conceptually important for the following reason. The ideal models that have been introduced in the curved space approach are three-dimensional curved templates. They look locally like three-dimensional Euclidean models. This is in contrast with the kind of ideal models considered in hyperspace cut and project schemes developed for quasicrystals, that are described here in appendix A7.

Since we shall need a rather detailed knowledge of this structure, called the polytope {3, 3, 5}, appendix A5 is fully devoted to its description (see also appendix A4), to which we urge the reader to refer as often as needed. Indeed, dealing with hyperspace figures is far from being trivial. We have tried to gather information of many different kinds: three-dimensional cuts, mapping, symmetry group, coordinates, etc . . . , in such a way that any interested reader can, with reasonable effort, build their own vision (either geometrical or analytical) of this polytope.

2.4 The {3, 3, 5} polytope: an ideal template for amorphous metals

2.4.1 The {3, 3, 5} polytope: the first curved space model

The {3, 3, 5} polytope was proposed as an ideal template for amorphous metals by one of us (J.-F. S.), and was first published in a common paper with M. Kléman (1979). The latter had readily recognized the important role that disclination lines would play in these curved space models, and subsequently focused on the consequences of building templates in a hyperbolic (negatively curved) space. The polytope approach (on S^3) was described in more detail in a subsequent paper (Sadoc 1981a).

So, the {3, 3, 5} polytope, which is a tiling by tetrahedra, provides a very dense atomic structure if atoms are located on its vertices. It is therefore naturally used as a template for amorphous metals, but one should not forget that it is at the price of successive idealizations where parts of the reality are lost.

- Most amorphous metals are alloys. We have chosen to describe an ideal monoatomic structure.
- Metallic atoms have complex electronic structures with different types of orbitals whose hybridization may play an important role. Here we have focused on isotropic 's' type of bonding. The atomic arrangement is coded into a sphere packing.
- Finally, last but not least, the template itself lives in an unphysical space (S^3).

It is interesting here to recall the numerical simulation work done by Straley (1986), who compares the relaxation of two structures towards their ground states. One is a disordered f.c.c. dense packing of atoms, the other is a set of 120 atoms interacting on S^3 (with a radius corresponding to that of the above polytope). Figure 2.5 displays energies plotted against the simulation time steps, the simulation containing annealing sequences. Clearly the Euclidean structure relaxes very slowly towards equilibrium – because of the frustration – while on S^3 the system easily finds its ground state, corresponding to the {3, 3, 5} polytope.

The richness of this approach will become obvious when we shall show that there exist deep intrinsic properties of the materials which survive these simplifications. We shall see further that it allows us to generate a general picture of amorphous solids, with qualitative and quantitative results, at geometrical and topological levels. The main result is probably to split the

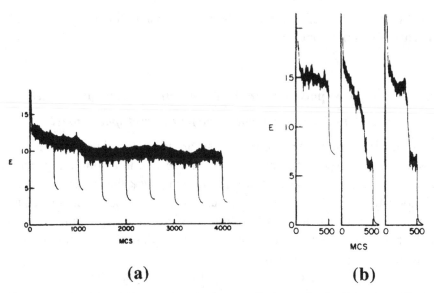

(a)   **(b)**

Fig. 2.5. Numerical simulations using the Monte-Carlo method, of the annealing of a previously disordered crystalline f.c.c. structure (a); compared (b) to the annealing of 120 atoms on S^3 (with appropriate radius). Only 500 steps (MCS) were made in order to obtain the polytope, since the Euclidean crystal is not obtained after several thousand steps. From J.P. Straley (1986).

structure into ordered regions and defects, as will be explained later. Let us first look at two simple quantities.

2.4.2 *The intersite distance distribution in the polytope*

The intersite distance distribution is easily calculated since, in a regular polytope, all the sites are equivalent and one can focus on the environment of the 'north' pole ($\omega = 0$ in the polar coordinate system of appendix A1, eq. A1.4). The geodesic distance to its neighbours is then simply given by $r\omega$. We know that the 119 other vertices are gathered into eight shells, which leads directly to the distance distribution function. At this stage, it is worth comparing the polytope distribution with experimental data. Figure 2.6 gives the radial distribution function of the polytope with the geodesic distance $r\omega$ scaled to the experimental values. Except for one far distance at about 11 Å there is an obvious one to one relation between the experimental data and theoretical distances, with slight metrical differences. This relation indicates (if not proves) that the medium range atomic arrangements in the amorphous material bear strong similarities to the polytope order.

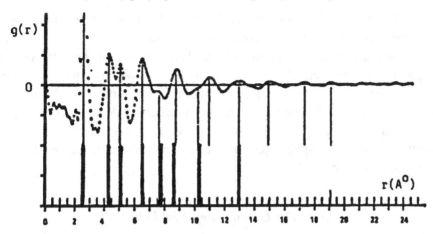

Fig. 2.6. Radial distribution function $[r(P(r) - 1)]$ obtained from X-ray diffraction on amorphous iron prepared by glow discharge (Lauriat 1983) compared to the {3, 3, 5} polytope distance distribution (delta functions in heavy lines and a radial distribution function obtained after mapping of the polytope).

2.4.3 The packing fraction of the polytope

In order to better understand this important question of compactness, it is interesting to compare the packing fraction of different structures. The packing fraction of the polytope can be easily calculated (Sadoc 1981b). One considers that hard spheres (balls) are placed at the polytope vertices with radius such they are in contact with neighbouring spheres. All the calculation is done in the spherical space S^3 (the hard spheres are 'curved' such that their interior belongs to S^3, and their curved radius is $r_1 = r\pi/10$). Let us first calculate the three-volume v of such a spherical 'cup' surrounding the north pole:

$$v = 2\pi r^3 \left(\omega_1 - \frac{\sin 2\omega_1}{2} \right) \tag{2.4.1}$$

Inserting the value $\omega_1 = \pi/10$, summing the 120 individual balls' contributions and dividing by the three-volume of S^3, one gets the filling factor

$$f = \frac{120v}{2\pi^2 r^3} = 0.774 \tag{2.4.2}$$

This is a very high value which exceeds the value 0.74 of the densest sphere packing in R^3 (realized for example by the f.c.c. or the h.c.p. packing). The reason is rather simple. The f.c.c. structure is a three-dimensional regular packing of tetrahedra and octahedra while polytope {3, 3, 5} contains only tetrahedra. The f.c.c. packing deficit is therefore due to the octahedra, which are less efficient packing configurations than tetrahedra.

Now, amorphous metals are known to be less dense than their crystalline counterparts. This apparent discrepancy between real systems and our model disappears when one realizes that a decurving procedure from S^3 to R^3 implies lowering the packing efficiency of the mapped structure. A simple estimation of the mapped structure density gives very reasonable values compared to both numerical simulation of sphere packings and experimental trends in density variations (see chapter 6, §2).

2.4.4 The Coxeter simplicial helix

A very interesting object, obtained by piling regular tetrahedra in one direction, bears some relation to the {3, 3, 5} polytope organization. Select two faces of a tetrahedron, on which two other tetrahedra are glued, and proceed in gluing new tetrahedra, with the condition that four tetrahedra never share an edge and such that edges with only one tetrahedron appear more or less aligned. A pseudo-linear set of tetrahedra is then obtained on which external edges form three helices (figure 2.7). Surprisingly, this set is not periodic owing to an incommensurability between the distances separating centres of neighbouring tetrahedra, and the pitch of the three helices (Coxeter

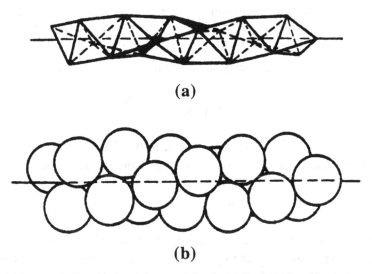

(a)

(b)

Fig. 2.7. A Coxeter helix obtained from a 'chaplet' of tetrahedra (a). A dense packing of spheres centred on the tetrahedron vertices (b).

1985, Boerdijk 1952). The vertices' coordinates take the following form (with unit edge length):

$$x_n = a \cos n\theta$$

$$y_n = a \sin n\theta$$

$$z_n = nc$$

$$\text{with} \quad a = 0.5196 \quad c = 0.3162 \quad \text{and} \quad \theta = 131.81 \tag{2.4.3}$$

Note that this is one of the simplest quasicrystalline structures that one can think of, with one single deterministic building rule! Several such helices are found in the polytope $\{3, 3, 5\}$, whose axes are great circles which belong to a discrete fibration of S^3 (forming a discrete Hopf fibration, see appendix A3). In R^3, the simplicial helix has pseudo-periods, for which the incommensurability is very low; for instance every 30 tetrahedra, the structure almost repeats itself in the x, y plane (indeed $30\theta = 10.96 \times 2\pi$). In the polytope, and owing to the space curvature, the set of 30 tetrahedra perfectly closes onto itself on a great circle.

2.5 Covalent tetracoordinated structures

The $\{3, 3, 5\}$ polytope is therefore an important template for metallic structures, but exactly as there are, in crystalline phases, many different structures which can be derived from the f.c.c. dense structure, the $\{3, 3, 5\}$ polytope gives rise to several interesting derived structures, some of which are described below.

2.5.1 Covalent disordered structures

This is the domain where the continuous random network, originally proposed by Zachariasen (1932), is the most commonly accepted model. A set of minimal rules is adopted among which the most important is that for each individual atom, the local coordination number is such that the $8 - N$ rule is satisfied (for atoms as columns 4, 5, 6 and 7). For example, a silicon atom (fourth column) is tetracoordinated while a phosphorus atom has three neighbours. Note that the continuous random network model is also used in the case of amorphous III-V and II-VI compounds where, as in their crystalline counterpart case, the atoms are tetracoordinated. This strict coordination rule is strongly supported by the experimental investigation: the pair distribution functions, obtained from the results of diffraction experiments, clearly show these coordinations (simple relations exist between first and second neighbour distances and orbital hybridization in covalent systems). SiO_2 glassy structures can also, to some extent, be described as decorated tetracoordinated structures, if we consider that oxygen atoms are on edges of a tetracoordinated graph; however, oxygen atoms introduce an additional flexibility to the network.

2.5.2 *Tetracoordinated structures*

Tetrahedrally coordinated covalent structures (diamond, silicon, germanium, silicates, etc.) form ordered structures based on hexagonal rings of bonds. In most cases, boat- or chair-like rings are present. Neither are planar and they both have equal bond lengths and tetrahedral angles ($109° 28'$) at their vertices, associated with sp^3 orbitals.

Configurations can also be characterized by the relative rotation of neighbouring tetrahedra along each bond. Eclipsed or staggered configurations (figure 2.8) are usually found in the ordered tetracoordinated structures. Consider three consecutive edges: the first two define a plane and so do the last two. The dihedral angle is the angle between these two planes.

The eclipsed configuration corresponds to a zero dihedral angle, the staggered one to an angle of $2\pi/6$. The 'chair' ring has all its bonds staggered, whereas the 'boat' ring has two eclipsed and four staggered bonds. These basic rings are the fundamental units in common periodic structures. Diamond, for example, consists only of chair rings while the wurtzite structure contains both

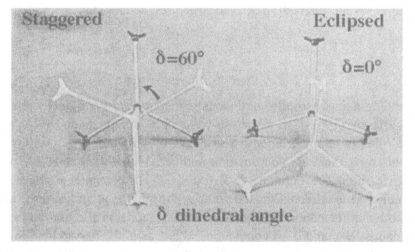

Fig. 2.8. The staggered and the eclipsed configurations. The dihedral angle is the angle formed by the two planes defined by an edge and its two adjacent edges. This angle is zero for the eclipsed configuration and $2\pi/6$ for the staggered configuration.

chairs and boats. The more complex structures of zeolites, like faujasite (Wells 1977), which contain larger rings surrounding empty channels, but also penta-gonal rings, provide other examples of tetracoordinated structures. Note that it is possible to show that, in all tetrahedrally bonded structures made of cages, the mean number of edges for rings is close to 5.1 (see the dual problem in appendix A4, §2.2).

In disordered covalent structures, like amorphous a-Si or a-Ge, short range order is found to be an almost perfect regular tetrahedral connectivity (Sadoc and Mosseri 1982a, b). But this local order can lead to numerous different structures in Euclidean or in curved spaces. It is therefore necessary to have information about the order at a scale larger than the first neighbour distance. The important topological ingredient in such local order description is the local ring configuration. Ring configuration is mainly characterized by the number of edges in the ring (and its parity), their possible twist and finally by the number of rings sharing a vertex or an edge. Now, in dense and tetracoordi-nated structures, the similarity of the local symmetry (tetrahedral interstices and tetrahedral binding) leads to geometrical relations between them. In curved space also, symmetry groups of both models have common sub-groups. This explains why we can obtain models for covalent structure starting from the $\{3, 3, 5\}$ polytope.

2.5.3 The polytope $\{5, 3, 3\}$

This is the polytope dual to polytope $\{3, 3, 5\}$ (observe in the formula $\{p, q, r\}$ the inversion between p and r). Geometrically it is obtained by centring the 600 tetrahedral cells of polytope $\{3, 3, 5\}$ and joining two such centres by an edge whenever their tetrahedral cells share a face. This polytope has 600 tetravalent vertices, 1200 edges, 720 faces and 120 cells. The duality is obvious in that the number of i-dimensional cells of the $\{3, 3, 5\}$ equals the number of $(3-i)$-dimensional cells of the $\{5, 3, 3\}$. Since each edge of the $\{3, 3, 5\}$ is shared by five tetrahedra, then each face of the $\{5, 3, 3\}$ is a pentagon and the cells are regular dodecahedra. In figure 2.9 are shown several cells of the polytope $\{5, 3, 3\}$ after mapping onto R^3, the outer dodecahedra already appearing quite distorted.

The $\{5, 3, 3\}$ and $\{3, 3, 5\}$ share the same symmetry group G (appendix A2, §3; see also A4 and A5). The $\{5, 3, 3\}$ vertices are the image under G of the orthoscheme vertex located at a $\{3, 3, 5\}$ tetrahedral cell centre. Now scaled to the first neighbour distance, the $\{5, 3, 3\}$ is less curved than the $\{3, 3, 5\}$. This is related to the smaller angular mismatch in R^3 when gluing three dodecahedra around a common edge compared to five tetrahedra. As a

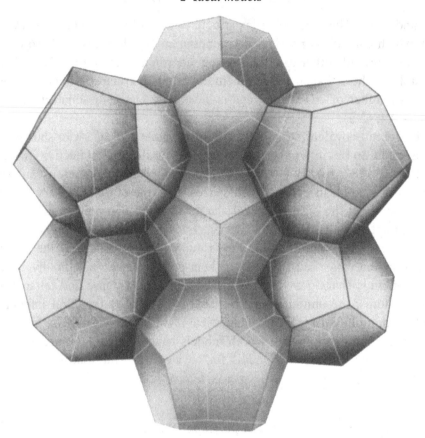

Fig. 2.9. Local arrangement of several dodecahedral cells of polytope $\{5, 3, 3\}$ after mapping on a tangent R^3 space.

consequence, with the same metrical distortion, it is possible to map more $\{5, 3, 3\}$ vertices than $\{3, 3, 5\}$ vertices. One can also say that the dodeca-hedral packing problem in R^3 is less frustrated than the tetrahedral packing.

The $\{5, 3, 3\}$ is a clathrate-like 'caged' structure. Indeed such dodecahedral cells are present in the clathrate crystal structure either in silicate compounds (Wells 1962) or in some ice structures (Davidson 1973). The polytope $\{5, 3, 3\}$ may be useful as a template for the non-crystalline versions of these materials. In fact we shall show below that even the crystalline clathrate structures (and their dual Frank–Kasper metallic phases) have a relationship with template polytopes and can be described as a polytope-like structure threaded by a periodic array of disclination lines.

As far as amorphous tetracoordinated semiconductors are concerned (like a-Si, a-Ge, a-GaAs, ...) the polytope $\{5, 3, 3\}$ does not seem to describe well their local order. Indeed the $\{5, 3, 3\}$ contains a very high number of pentagons, whose presence leads to definite intersite distances which are not experimentally observed (at least significantly). Furthermore, the expected density, with $\{5, 3, 3\}$ as template, is much lower than the measured one. These materials are much better described in relation to polytope '240' which is presented below.

2.5.4 The polytope '240'

Let us now describe this polytope, which has in some respects a more subtle structure, and shares only parts of the $\{3, 3, 5\}$ symmetries, even though it can be described as a decorated version of the latter. Let us follow a building rule similar to that which leads to the diamond structure starting from the f.c.c. structure: a new vertex is placed at the centre of some tetrahedral cells of the compact structure. In the f.c.c. case, one tetrahedron over two is centred, while in the present case, one tetrahedron over five will be centred, which has the consequence of breaking the five-fold symmetry of the polytope $\{3, 3, 5\}$, only a ten-fold screw axis being preserved. One gets a regular structure with 240 vertices, called polytope '240', which is chiral: it cannot be superimposed on its mirror image. Its direct symmetry group is presented in appendix A2.

Another way to describe the '240' (it is in fact the way we first generated it), is to recall that the diamond crystalline structure can be obtained by starting from an f.c.c. structure and adding a second replica of the f.c.c. structure, translated by $(1/4, 1/4, 1/4)$ with respect to the first one. Similarly, polytope '240' is generated by adding two replicas of the $\{3, 3, 5\}$, displaced along a screw axis of S^3. It is amusing to note that, while we thought we had 'discovered' this fascinating structure in 1981, Coxeter finally remembered that it had already been described half a century before by Robinson (1931)!

The '240' is not a regular polytope in the Coxeter sense, and so cannot be denoted by a triplet $\{p, q, r\}$. A three-dimensional graph made of vertices and edges is not necessarily a polyhedral packing: a typical example of such an impossibility is provided by the diamond structure. It becomes possible only if any pair of edges defines a face shared by two polyhedral cells. The polytope '240' is also an example of a graph (embedded here in S^3) where one encounters ambiguities while trying to define faces and cells. For this reason, even the term 'polytope' is not well adapted to denote this structure, but we keep it to recall that it is an ordered structure on a hypersphere.

Let us look at its structure in more detail. Each vertex of the first $\{3, 3, 5\}$ replica is surrounded by four vertices from the second replica. The polytope is then an alternated, bicolour structure and contains only even-membered cycles; the smallest are hexagons in a twisted boat configuration which brings the dihedral angle into a value intermediate between the eclipsed and the staggered case without modifying the 'bond' angles and the first neighbour distances (this is possible owing to the S^3 space curvature).

The local configuration is perfectly tetrahedral as can be seen in figure 2.10 which shows orthogonal mapping of subsets of increasing size of the polytope. Three perfect boat rings can be arranged such as to form an eight-vertex cluster called the 'small barrelan' (figure 2.11), which is a building block of the wurtzite crystalline structure. The twisted hexagons of the '240' are also arranged by three such as to give twisted small barrelans, the twist being along the barrelan three-fold axis, the latter being threaded by 'channels' somewhat similar to those present in the diamond or wurtzite crystals. These channels follow great circles of S^3 and are bounded by a triple arrangement of chains of sites (a triple helix) close to the (1, 1, 0) chains of the diamond structure.

We have not described yet the frustration problem to which the polytope '240' provides an answer. Suppose that one tries to construct a tetravalent cluster in R^3 in such a way as to maximize the interconnectedness of the structure, e.g. the number of rings sharing a common vertex. Looking to

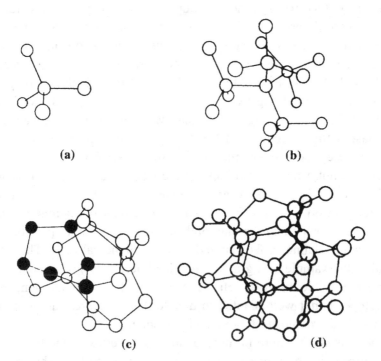

(a)   (b)

(c)   (d)

Fig. 2.10. Local view of the polytope '240'. (a) Five vertices, the tetrahedral symmetry is clearly visible; (b) 17 vertices; (c) 21 vertices, a six-fold ring is distinguished; (d) 39 vertices.

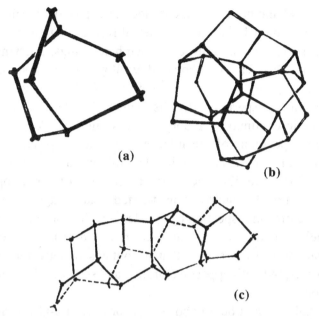

Fig. 2.11. With the polytope '240' it is possible to have as many as 18 rings passing through a vertex. (a) A small barrelan. (b) An assembly of four small barrelans sharing a central site and forming a 27-vertex cluster. This four-valent configuration presents a high degree of interconnectedness. (c) The local bonding arrangements of sites around the 30/11 channelling axis.

structures with hexagons, the diamond and wurtzite crystalline structures have equivalent interconnectedness: twelve hexagons through each point.

In fact these two structures, in terms of graphs, are equivalent to a much higher degree: the numbers of returning walks of n steps in the network for any value of n, are the same. In a simple tight binding model for elementary excitations, they are 'isospectral'.

Now, at the price of a limited distortion, it is locally possible to put more hexagons sharing a common vertex. By twisting boat-like hexagons, it is possible to have as many as 18 rings passing through a vertex. This configuration cannot however propagate freely in R^3; it is geometrically frustrated. But it can be realized on S^3 in the form of the polytope '240'. Note that this local order minimizes a tight binding form of the energy for small clusters. Indeed the number of dangling bonds (which cost energy) per atom is lower than for small clusters of diamond or wurtzite structure. Finally, let us stress that polytope '240' is locally denser than the diamond structure. This is exactly the corollary of what was said above for the $\{3, 3, 5\}$ compared to the f.c.c. dense structure.

Another interesting fact is that polytope '240' has a very close relationship with a well known hand built tetravalent model, the so-called Connell–Temkin (1974) model. This model plays a very special role among continuous random networks because it contains only even-membered rings, allowing us to describe suitable models for amorphous III-V alloys, where a strong chemical order imposes even parity on the ring. Dixmier *et al.* (1984) have given a rather decisive argument to experimentally distinguish the Connell–Temkin model from more generic continuous random networks like the Polk (1971) model, involving relative positions of the interference function peak. The Connell–Temkin model is an extreme case in the class of continuous random networks, and is strongly related to polytope '240'. Indeed as in the polytope case, the Connell–Temkin model contains many twisted boat-hexagons forming small barrelans. Even more, an inspection by eye done by Sachdev and Nelson, of the original Connell–Temkin model (found covered by dust in a forgotten part of the Harvard laboratory) has shown that it contains regions, modelled on the '240' type of order, but with opposite chirality, and separated by a thin interface defective region.

If polytope '240' should be a good model for a-Si or a-GaAs, an important question is whether the intrinsic local chirality can be observed. However there are several experimental difficulties. It is for instance possible that the actual material contains regions of opposite chirality, with the same amount, leading to an overall racemic material, making this characteristic therefore difficult to detect. Nevertheless, it has been suggested that a third harmonic generation in optical experiments may be sensitive to the local chirality (DiVincenzo 1988b).

2.5.5 Chirality in three-dimensional networks with hexagonal rings

We have seen that it is possible to twist hexagonal boats, leading to a local chirality. Let us try to propagate this type of order, in a way slightly different from that of polytope '240' (Major *et al.* 1987). We start by building the network around a cylinder. Two types of rings are present: those which lie on the surface of the cylinder and those surrounding its principal axis, which are all taken to be hexagons. There are essentially two elementary ways to identify opposite sides in order to obtain cylinders as described in figure 2.12.

The configuration shown in figure 2.12a corresponds to wurtzite, with chairs encircling the channel axis, and boats on the surface of the cylinder.

Figure 2.12b shows a chiral structure in which all rings are twisted boats. It is very close to what is found in the polytope '240'. Note that the cylinder section is smaller in this case (2.5 hexagons) than in the wurtzite (3 hexagons). This structure is locally denser. But it is impossible to extend it to

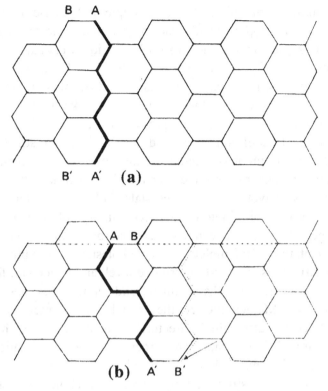

Fig. 2.12. (a) Cylindrical channels in wurtzite formed when A is identified with A', B with B' etc. The bold lines form a ring surrounding the cylinder axis. The structure is non-chiral. (b) Structure formed by removing the region above the dashed line from the cylinder. Again, corresponding letters are identified. The resulting structure is chiral.

the entire Euclidean space: only its curved version (the polytope '240') is unfrustrated.

The Coxeter simplicial helix was presented in §2.4.4. The above cylinder is in fact a decoration of this helix. Note that the above considerations might find new applications in the light of the recently found carbon 'tubullene' structures.

2.6 Frustration in lamellar liquid crystals and amphiphiles

2.6.1 Liquid crystals

A fluid is classically considered as an assembly of molecules whose degrees of freedom, positions, orientations and velocities, are more or less disordered, but not randomly distributed like in a gas. There exist, however, situations in which the fluid is periodically modulated as, for example, in the case of lamellar and smectic phases of lyotropic and thermotropic liquid crystals, and in chiral cholesteric phases (de Gennes and Prost 1993). Liquid crystals are observed mainly by optical techniques, as done originally by O. Lehmann in 1889. He gave the names to these new phases, impressed by their anisotropic properties, although, at a microscopic level, they clearly behave like liquids, even if their viscosity is sometimes great, like in a grease. George Friedel (1922) studied these phases in great detail and introduced the classification, still in use, with the concept of mesomorph phases and the terms 'smectics' and 'nematics'.

Stratification can concern the molecular position, as in lamellar phases and the smectic phases of lyotropic liquid crystals ('alloys' of water and organic molecules) or thermotropic liquid crystals (containing only organic molecules with an intermediate state between solid and isotropic liquid); but it can also concern the orientation of the molecules as in cholesteric chiral blue phases. To describe these structures, at a first descriptive level, and to account for some of their properties, we can forget the chemistry and suppose that these molecules are like small rods. Mathematically, they can be represented as undirected (headless) unit vectors, also called 'directors'. In nematic phases, the latter are more or less parallel, but without any order as far as the position of their centres is concerned (figure 2.13). Smectic phases can still be thought of as small parallel rods, but structured into layers. A cholesteric structure is schematized as a continuous organization of stratified fluids in flat layers, where molecular axes are parallel inside a layer, but their orientation rotates continuously from a layer to an other. The pitch of the cholesteric is defined as

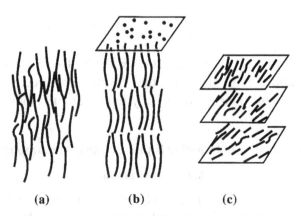

(a)   (b)   (c)

Fig. 2.13. Schematic representation of molecules in a nematic (a), a smectic (b), and a cholesteric (c) structure.

double the distance between two layers having parallel orientation (in order to have an overall 2π rotation).

In the next section we detail two examples of frustration in liquid crystals: in lyotropic crystals formed from alternating layers of water and organic molecules, and in blue phases, which are systems related to cholesterics but in which the twist of the small rods is not along a single axis.

The most observed situation and the easiest to understand concerns organizations of stratified fluids in flat layers, periodically stacked along one direction, which result from the constant distance imposed by the interactions between interfaces or, in the cholesteric case, from a constant pitch of the chiral twist.

However, when some thermodynamical parameter is varied, the systems often build more complex organizations, periodic along two or three directions, for example in hexagonal and cubic phases or in blue phases. We shall see that these new organizations, found in rather different systems, result from a frustration effect, arising from the conflictual requirement of constant distance and the appearance of a new symmetry in the stratifications, introduced by a lateral area difference between different layers of the stratification or by a so-called 'double twist' condition in the chiral case. The frustration appears related to the flatness of the embedding Euclidean space, but it can be relaxed if the embedding space is given an adequate curvature.

2.6.2 Frustration in crystals of films

The characteristic scale for this structuration is not related to the constituting molecules themselves, but to the characteristic length scales of the continuous films arranged periodically along directions normal to the layers. This kind of structure has been extensively studied by Luzzati (1968) who was one of the first to understand its complexity: it behaves like a liquid structure at the molecular scale, although the order is crystalline at a larger scale.

As said before, in the case of lamellar and smectic systems the stratification concerns the positions of the molecules. We analyse the problem in the case of systems of amphiphiles and lamellar phases only, but similar considerations hold for mesogenic molecules and smectic phases.

The film, the bi-layer, is made of two identical layers and several forces fix the inter- and intra-layers distances: van der Waals forces, electrostatic interactions between polar groups at the interfaces, forces created by water polarization and charge distributions in aqueous layers, hydrophobic interactions which prevent the presence of water within the layers, and fluctuation-induced forces of entropic character.

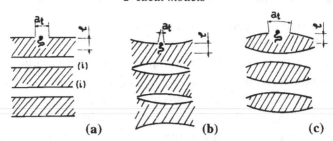

Fig. 2.14. Schematic representation of a periodic system of films with flat interfaces: (a) constant interfacial distances and zero curvature are compatible in R^3; (b) and (c) constant interfacial distances and non-zero curvatures are no longer compatible in R^3 and the system becomes frustrated.

The precise roles of these different interactions, as well as their mutual interplay, are not fully understood at the moment. However, it is reasonable to state that the main effect of the components normal to the interfaces is to maintain a constant distance between them, if the interfaces are homogeneous, while the parallel components determine the interfacial curvatures. Owing to the symmetry of the film with respect to its middle surface, it is clear that imposing symmetric curvatures for the two facing interfaces is not compatible with keeping a constant distance between the interfaces, if a lamellar type of stacking is kept, as shown in figure 2.14. The flat interfaces case is represented in figure 2.14a, while figure 2.14b shows one conflictual situation with curved interfaces. This is a typical case of frustration which has no solution in the three-dimensional Euclidean space R^3. As a consequence, the system will adopt a more complex structure, in order to minimize the energy, that is by building the best compromise between these conflictual forces.

2.7 Lamellar structures in curved spaces

Having recognized a new case of geometrical frustration, it is natural to try to solve the problem, as in the case of tetrahedral packing, by allowing the embedding space to be curved. This is indeed possible and such ideal models have been proposed for bi-layers in three-dimensional curved spaces (Charvolin and Sadoc 1987, 1988a).

We limit this first approach to curved spaces with homogeneous positive curvatures, because we look for configurations which relax the frustrations equally everywhere; the positive curvature allows us to replace the condition of periodicity by a cyclic condition. We have two possible choices in three dimensions: the hypercylinder $S^2 \times R^1$ and the hypersphere S^3. Some of their properties, which are needed in the following discussion, are described in appendix A1.

Facing the difficulty of pictorial representations of three-dimensional curved spaces, embedded in R^4, which is only practically solved by the use of stereographic projections onto R^3, we find it useful to discuss first a simple two-dimensional problem of frustration in the Euclidean plane R^2, which is relaxed on the ordinary sphere S^2. Recall that we proceeded along the same line with discrete systems, describing frustration in a two-dimensional pentagonal arrangement before treating the three-dimensional tetrahedral packing problem.

2.7.1 A simple two-dimensional example

We just consider the intersection of a three-dimensional system of layers by a plane R^2 normal to the layers. Doing so, we have a two-dimensional model where a film becomes a strip. As said above, there is no frustration when the interfaces limiting the film are flat, their two-dimensional images being two parallel straight lines bordering a strip. This results from the fact that the interfacial and middle lengths are equal, corresponding to equal equilibrium distances for molecules in the bulk of the film, and molecule heads on interfaces. Frustration appears when the interfaces become symmetrically curved, or when the interfacial and middle lengths become different, as a consequence of two different equilibrium lengths between hydrophobic parts of the molecules in the bulk and hydrophilic parts at the interface. In our two-dimensional case, this situation can be homogeneously relaxed when the flat space R^2 is curved into the spherical space S^2 of radius R if, as shown in figure 2.15, the middle line is placed on the equator and the shorter interfacial lines are placed on 'parallels' at equal distances from the equator. The frustration is obviously suppressed, as the interfacial lengths $2\pi R \sin \theta$ are smaller than the

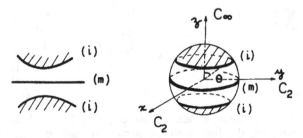

Fig. 2.15. Example in two dimensions of a frustrated strip, because its middle length is different from the edge length, forbidding parallelism; the frustration is eliminated by transferring the strip onto S^2.

middle length $2\pi R$, while the basic symmetry of the system is preserved, as the equator is a stationary line separating the sphere into two identical hemi-spheres. Moreover, when moving along meridians, which are great circles and therefore geodesics of S^2, the interfaces of the strip are periodically crossed, as when moving along the normals to the interfaces of a lamellar system. This finite system, with one strip in the curved finite space S^2, is therefore equivalent to the infinite periodic system of strips, but here without frustration. Finally, the symmetry axes of the relaxed structure in S^2 are: one continuous axis normal to the equatorial plane and a continuous family of two-fold axes in this plane.

2.7.2 Lamellar and smectic systems in $S^2 \times R^1$ and S^3

Frustration arises from area differences between the middle surface and the interfaces of the films. We proceed in a quite similar way to that in the above two-dimensional example, but with the use of three-dimensional curved spaces. The frustration can be homogeneously relaxed by curving R^3 into the cylind-rical space $S^2 \times R^1$ or the spherical space S^3. The middle surface of the film is placed on a particular surface of the curved space which divides it into two equal subspaces and is a stationary surface. In the case of $S^2 \times R^1$, such a surface is the usual cylinder $S^1 \times R^1$. In the case of S^3 there are two possible surfaces: the great sphere S^2 of S^3 and the so called spherical torus T_2. These surfaces are shown in figure 2.16 using a stereographic projection of the curved spaces onto R^3; they are also described with more details in appendix A1, §2.2 and §2.3.

The two interfaces are placed on equidistant, parallel cylinders, tori or spheres. In these situations the symmetry of the film is preserved, as the

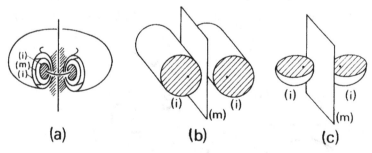

(a)   (b)   (c)

Fig. 2.16. The three-dimensional periodic system of parallel films. Frustration in R^3 is suppressed by transferring the film onto the particular surface which separates the curved spaces $S^2 \times R^1$ (a) and S^3 (b, c) into two equivalent subspaces. The periodic character is replaced by a cyclic one in the curved spaces. (Stereographic projections of the spherical component, (i) means interface and (m) means middle surface).

surfaces supporting its middle surface are stationary surfaces separating the spaces into two equal subspaces; the frustration is relaxed, as the interfacial areas are smaller than the middle area, and the distance between the interfaces is kept constant. The periodicity of the stacking is replaced by a continuous cyclic displacement along geodesics normal to the surfaces. Finally, the symmetry axes of relaxed structures can be described. There are two-fold as well as continuous axes. The first are, in $S^2 \times R^1$, great circles normal to $S^1 \times R^1$ and its generators, in S^3 geodesic great circles normal to T_2 or in the surfaces of T_2 and S^2. The second are the polar lines of the spaces.

2.8 Frustration and curved space structure for blue phases

2.8.1 Frustration in cholesteric systems

Chiral molecules lack inversion symmetry; they have a 'handedness' and are distinguishable from their mirror image. A pair of neighbouring molecules will have an equilibrium position corresponding to a minimal energy if they sit at a slight angle with respect to each other (imagine close 'screws'). This is the case of cholesteric molecules which are characterized by a director and a natural twist. It is then more convenient to leave the molecular scale, and to consider only averaged molecular orientation in a small region of the space. The structure is then described, in this continuous approximation, by a field of directors, but, except for description and classification of topological defects, it is possible to consider a vector field, just forgetting the vector orientation.

The classical cholesteric phases, described in figure 2.17, correspond to the case where the twist propagates along one direction only. There is a stratification of the orientation of the director, which is subjected to a twist, so that a period appears corresponding to half the pitch of the helicoidal reorientation. But a more complex situation arises if one allows for an isotropic propagation of the twist along all directions perpendicular to the molecular axis, which is called a 'double twist' configuration. A schematic drawing of this situation is presented in figure 2.18. Such a situation is frustrated as can be understood in figure 2.19, by following and keeping track of the director orientation along two different paths ending at the same point. It is then impossible to fill the space with a vector field which obeys everywhere the 'double twist' rule. This is therefore a new example of geometrical frustration, which can be expressed here in a differential geometry language:

$$\partial_i n_j = -q\epsilon_{ijk} n_k \qquad (2.8.1)$$

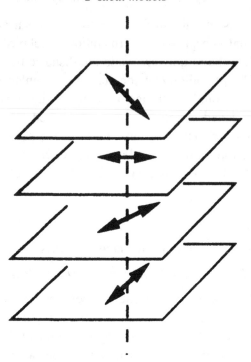

Fig. 2.17. Schematic representation of a cholesteric system with a twist along one direction only, this simple twist and the layering being compatible in R^3.

where n_i is the component of the field, $2\pi/q$ is the cholesteric pitch and ϵ_{ijk} is a totally antisymmetric tensor in its indices with $\epsilon_{123} = 1$.

At low temperature, the chiral nematics mainly show a cholesteric structure whose coordinates read:

$$n(x, y, z) = (\cos qz, \sin qz, 0) \qquad (2.8.2)$$

This phase rotates only in one direction. It is also possible to minimize the 'double twist' energy in the central part of a cylinder, called 'double twist tube' with the field:

$$n(r, \theta, z) = \mathbf{u}_z \cos qr - \mathbf{u}_\theta \sin qr \qquad (2.8.3)$$

It has been shown that, in an isolated 'double twist tube', the structure decreases its energy by adopting a lower pitch than the natural cholesteric one (Dandoloff and Mosseri 1987). The classical models for blue phases are described as a tridimensional periodic arrangement of these tubes (Meiboom *et*

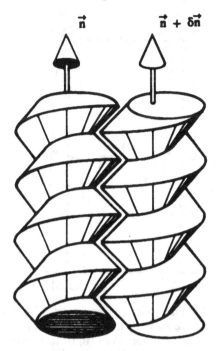

Fig. 2.18. Schematic representation of two chiral cholesteric molecules by screws, showing how they are aligned with a slight angle.

al. 1983, Hornreich *et al.* 1982). However, as will be shown in a later chapter, disclinations are necessarily present.

2.8.2 Cholesteric systems in S^3

This fascinating example of geometrical frustration was the first to be treated in a continuous system, by Sethna *et al.* (1983), using an ideal template in S^3 – later, however, than the discrete ideal models with polytopes. These authors have shown that a vector field tangent to the great circles of a Hopf fibration (see appendix A3) is a field with a 'double twist' without frustration, if the radius of S^3 is related to the pitch of the cholesteric in an appropriate way. One analytical possible form is

$$n(x) = q(-x_1, x_0, x_3, -x_2) \qquad (2.8.4)$$

More precisely (Pansu *et al.* 1987), the 'double twist' condition is related to the Levi–Civita connection on S^3 (see appendix A8) and can be written

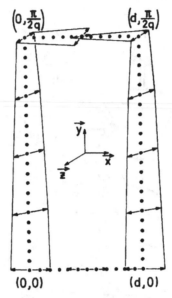

Fig. 2.19. Frustration in a 'double twist' system: start with a director at the origin pointing in the x direction. As we move in the y direction it rotates, until it lies in the z direction. As we move in the x direction from $(0, 0)$ to $(d, 0)$ it remains parallel. At $(d, 2\pi/q)$ it points in the z direction. Finally, frustration is clearly identified when the two travelling paths meet at point $M = (d/2, \pi/2q)$, and the vectors are differently orientated.

$$\partial_i n_j = \left(\frac{1}{r} - q\right)\epsilon_{ijk} n_k \qquad (2.8.5)$$

Then, if $r = 1/q$, the field is 'parallel' for this connection, which suppresses the frustration. This 'Hopf' field can be visualized, using the stereographic projection on R^3, as shown in figure 2.20a.

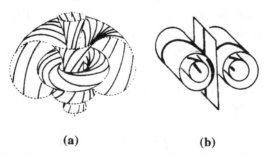

(a)   (b)

Fig. 2.20. The cholesteric system with 'double twist'; suppression of the frustration in R^3 with fibrations of (a) S^3, (b) $S^2 \times R^1$ (both represented by a stereographic projection onto R^3).

In figure 2.20b a field having a 'double twist' is drawn in the space $S^2 \times R^1$ (see appendix A1, §3): this solution is not as efficient as the S^3 solution because the twist pitch is not constant. It is an intermediate solution between the 'double twist tube' frustrated case in R^3 and the unfrustrated S^3 Hopf field solution.

2.9 Frustration in polymers

Following the line of thought of the preceding model for the blue phase, Kléman (1985a, b, 1989a, b) studied the problem of disordered polymers, which present both discrete aspects (polymers are single individual objects), and continuous aspects (each one is represented by an infinite line in space). Such a system is governed by two kinds of forces: attractive forces which lead to a dense system, and entropic forces which tend to bend the molecules. There is competition between these two geometrically opposite effects: a tendency towards an equal distance between chains of polymers, and a tendency toward misorientation. Under this competition, the local state looks like a twisted strand.

The helicoidal winding of two chains is similar to the geometry of a 'double twist tube' in blue phases. Equidistant lines representing polymer molecules are along the force lines of the director field. Kléman proposed building an ideal model using a discrete sub-set of the Hopf fibration (see appendix A3). In fact, he treated the problem using a projective space representation, but we prefer to rephrase his model in the formalism used in this book.

It is possible to organize, in a dense and homogeneous way, sets of lines (which are great circles of S^3) having everywhere two, three, four or five neighbouring chains. The study of the decurving process towards the physical space R^3 and of the resulting topological defects has not yet been done.

There is another possible way to model the competition between attractive and entropic forces, using a negatively curved space, namely the hyperbolic space. This choice is driven by the fact that the two notions of parallelism and equidistance between two lines are disconnected in this space (see figure A1.1). Two lines at constant distance lose their orientational correlation. But two parallel geodesic lines are not at constant distance. Consider a model of a polymer in which chains are along lines at constant distance in a hyperbolic space. The loss of the angular correlation between two close lines increases with the space (absolute) curvature. In a Euclidean space, the decorrelation between two neighbouring chains increases with the entropic terms. In this view, defects necessary to decurve the hyperbolic space would be associated to the entropy. We are therefore facing a frustration effect resulting from both entropy and energy requirements, that flat space geometry cannot solve. Note that the particular properties of hyperbolic geometry have already been used in order to model chaotic motion (Balazs and Voros 1986).

3

Finite structures

3.1 Finite clusters

There is much experimental evidence that atomic finite clusters do not present necessarily the same geometrical order as in their crystalline counterparts. For instance, rare gas clusters often display icosahedral symmetry, their detailed structure depending on atomic species and size (Farge *et al.* 1975). It is now possible to grow and study metallic or covalent clusters, ranging from a few atoms up to several hundred atoms (Joyes 1990, Paillard *et al.* 1994). They are interesting not only for themselves (in the context of catalysis for example) but also to better understand cohesive properties of solids, or in the context of amorphous structures, since they can possibly model the local order of the non-crystalline solid during the growth process. One should not forget however that surface effects play a dominant role in cluster stability while their role in glasses should only be invoked for a dynamical process.

Several types of cluster have been proposed in the past. Some of them have a close relationship with mapped polytopes. The main results have been summarized elsewhere (Mosseri 1988a, Mosseri and Sadoc 1989), which we now describe.

3.1.1 Cluster indexation

We want to index finite clusters which can be derived from a given polytope P whose sites are decorated by atoms. A finite portion of P is mapped onto a tangent hyperplane R^3. We need to specify the tangent point T and the polar angle ω which limits the region to be mapped, and as a consequence the size of the mapped cluster. This is shown schematically in figure 3.1. Metrical properties depend on the precise location of the projection point Q which is defined

45

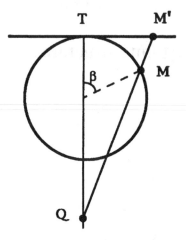

Fig. 3.1. Partial mapping. The polytope is embedded in S^3. M (with polar angle β) is mapped onto M'. T is the tangency point; Q is the projection point and d is the distance TQ, defined in the text.

by its distance d from T. Note that this position may not necessarily be unique for all the values of β ($\omega = \max(\beta)$). The general notation for the cluster is

$$C = [P, T, \omega, d] \qquad (3.1.1)$$

The first three quantities completely define the cluster from a topological point of view, while the last one is of a metrical nature.

3.1.2 Compact clusters: the polytope {3, 3, 5} and icosahedral local order

Upon mapping the {3, 3, 5} polytope, clusters with a perfect icosahedral order are obtained; note that other local orders (for example tetrahedral) can also result according to the polytope orientation (rotation of the embedding space S^3 before mapping). The price to pay is that the clusters are fairly small (recall that the polytope has 120 vertices). Table 3.1 gives a description of these clusters as well as their energy per site calculated within a simple Lennard-Jones potential scaled such that its minimal value occurs for $r = 1$:

$$V(r) = \left(\frac{1}{r^{12}} - \frac{2}{r^6} \right) \qquad (3.1.2)$$

E_1 corresponds to a unique d optimized for each cluster as a whole, while, for E_2, d is shell-dependent. The cases $\omega \geqslant 2\pi/3$ are not considered further because they contain unrealistic intersite distances. The first four clusters,

Table 3.1. *Clusters mapped from the
{3, 3, 5} polytope and their energy per
site. The absolute value of the minimum
of the Lennard-Jones potential is taken
as unity*

$\beta(rd)$	N	E_1	E_2
$\pi/5$	13	−3.41	−3.41
$\pi/3$	33	−4.28	−4.32
$2\pi/5$	45	−4.15	−4.5
$\pi/2$	75	−4.16	−4.54
$2\pi/3$	87	−3.06	−4.22
$3\pi/5$	107	−3	−4.3

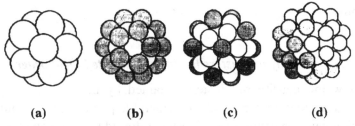

(a)   (b)   (c)   (d)

Fig. 3.2. The first four clusters ($N = 13, 33, 45, 75$) obtained by mapping the {3, 3, 5} polytope.

shown in figure 3.2, are very familiar in this field and have also been used to model the growth of quasicrystals (Romeu 1988).

3.1.3 Geodesic hyperdomes

Regular triangulations of S^2 based on the icosahedron have been popularized by Buckminster Fuller and are called geodesic domes. It is possible to generalize this idea for higher-dimensional spheres and here to S^3 (Sadoc and Mosseri 1985a). The {3, 3, 5} tetrahedral cells are iteratively filled by larger and larger portions of f.c.c. lattice. We call these successive polytopes G_i (G_0 being the {3, 3, 5}). Table 3.2 describes clusters mapped from G_1 and their corresponding energies. The 55 atom cluster is often called the 'Mackay icosahedron' in the field of quasicrystals (Mackay 1962, 1982). These clusters are observed experimentally, for instance when a beam of argon atoms produces free clusters by condensation in a vacuum chamber. Farge and coworkers (1975) have

Table 3.2. *Clusters mapped from the G_1
polytope and their energy per site*

$a(rd)$	N	E_1
$\pi/10$	13	−3.37
0.56	43	−4.45
$\pi/5$	55	−4.83
0.82	115	−5.36
$3\pi/10$	127	−5.19
1.02	157	−5.26
$3\pi/3$	177	−5.32
1.14	237	−5.1

observed such clusters of increasing size, among which for instance is the one
with 55 atoms.

3.1.4 *Density of clusters having a frustrated local order*

Consider now the density of clusters obtained by mapping polytopes. Sire
(1990) has shown that, within relatively general hypotheses, one should expect
a compression near the centre of the cluster followed by a dilatation towards its
surface. Let us summarize this calculation. Instead of considering a set of n
atoms on S^3 interacting for instance through a simple pair potential, let us
simplify further the study by supposing a continuous and uniform density

$$\rho = \frac{n}{\pi(2\omega - \sin(2\omega))} \tag{3.1.3}$$

where the denominator is the (three-dimensional) volume of the hyper-spher-
ical cup defined by the angle ω on S^3. Let $f_\omega(\beta)$ be the distance from the centre
of the cluster, after mapping of the shell defined by the angle β ($\beta \leqslant \omega$). On
S^3, the structure is supposed free of frustration; the energy density is minimal.
Decompose the energy of a shell into a tangential part, sensitive to local strain
through the ratio $f_\omega(\beta)/\sin\beta$, and a radial part, which depends on the derivative
of $f_\omega(\beta)$ with respect to β. After mapping, the energy of the cluster is:

$$E(\omega, f_\omega) = \int_0^\omega (V(f_\omega(\beta)/\sin\beta) + V(f'_\omega(\beta)))2\pi(\sin\beta)^2 \, d\beta \tag{3.1.4}$$

The most general solution to this integral is not easy to get, so we shall try a
solution of the form

$$f_\omega(\beta) = a_\omega^{-1/6}\beta, \quad c_\omega = a_\omega^{1/6}$$

Optimization with respect to a_ω leads to

$$f'_\omega(\beta) = c_\omega^{-1} = 1 - \frac{\omega^2}{20} + o(\omega^2)$$

The cluster is therefore compressed near its centre. Let $n(r)$ be the number of atoms in the cluster at the distance r, one finds

$$\frac{n(r)}{n} = \frac{2rc_\omega - \sin(2rc_\omega)}{2\omega - \sin(2\omega)} \tag{3.1.5}$$

This is indeed very close to what is observed by numerical simulation of clusters.

3.1.5 Improving the Mackay-like clusters' compactness

Regarding the importance of the Mackay icosahedron in quasicrystals, we would like to mention here an interesting fact about its compactness (Sire and Mosseri, 1998). We have seen above that it can be obtained by mapping a limited part of the 840 vertex polytope G_1. Coxeter has proved (Coxeter, private communication) that the compactness of the latter can be slightly improved by suitably moving some of its vertices. As a result one should expect that the clusters mapped from this modified polytope also have a larger filling factor. And indeed this is found to be true, even though the improvement is minute.

But, perhaps even more interestingly, all the mapped clusters are now chiral, with complex distortions as compared to the rather simple displacements that are required in the template polytope to make it more compact. We believe that the finding of these optimized clusters is an argument in favour of a synthetic, geometrical approach. We left it as a challenge to specialists of numerical simulations to get these denser clusters directly in R^3, starting from the symmetrical clusters. Indeed the optimal displacements are collective and cannot be simply achieved by moving the sites one at a time along a path on which the density is continuously improved. The fact that the deformed polytope is optimal in compactness does not prove that our mapped clusters are optimal. And indeed we are inclined to believe that their compactness could again be iteratively improved, but with smaller and smaller displacements and gains in compactness.

The next natural question is whether this greater compactness will stabilize the clusters. It was our initial hope that it would indeed be the case under standard pair potentials (Morse or Lennard-Jones). This would make this

example a very nice classical analogue of the quantum symmetry breaking
stabilization due to the Jahn–Teller effect. It is however not the case, as was
tested numerically. The reason is a subtle one. The compactness is affected
only by first neighbours' repulsive interactions. The energy is obtained as the
sum of pair interactions between sites which can be farther away. The gain that
some close distances bring is more than balanced by the loss from other pair
distances, which play no role in compactness.

3.1.6 Tetravalent clusters: the polytope {5, 3, 3} and dodecahedral local order

Almost 40 years ago, Tilton (1957) described dodecahedral packings of increasing size. He noted that, with one central dodecahedron, the more symmetrical clusters contain 13 and 45 units. If we call C the centre of a $\{5, 3, 3\}$
cell, they correspond to:

$$[\{5, 3, 3\}, C, 0.388rd, d]: 1 \quad \text{dodecahedron}$$

$$[\{5, 3, 3\}, C, 0.962rd, d]: 13 \quad \text{dodecahedra} \qquad (3.1.6)$$

$$[\{5, 3, 3\}, C, 1.435rd, d]: 45 \quad \text{dodecahedra}$$

d is not known, which means that our mapped clusters are only topologically
equivalent to those built by Tilton. He also described a cluster centred on a
vertex with 54 dodecahedra. The latter is equivalent to $[\{5, 3, 3\}, V, 1.475rd,$
$d]$ where V is a vertex of the $\{5, 3, 3\}$. Note that none of the clusters have
$\omega > \pi/2$. Tilton noted that larger clusters would have too much strain, which is
evident from the the polytope point of view since the next shell's pre-image
would belong to the S^3 south hemi(hyper)-sphere. A more metrical comparison
can be done with the work of Coleman and Thomas (1957) who give not
only the number of sites but also the distances from a central vertex. Their
dodecahedral cluster corresponds to $[\{5, 3, 3\}, V, \pi/3, d]$. Table 3.3 compares their data with those calculated for the polytope (with geodesic distances
scaled to a first neighbour of 2.35, appropriate to silicon atoms). The chosen
value of d is $2.75R$. The small differences between the hand-built and
polytope-mapped models simply mean that one should not use a constant d
value to fit the data.

3.1.7 Tetravalent clusters: decorated geodesic hyperdomes

It is possible to decorate the above f.c.c. regions in the G_i polytopes so as
to give diamond lattice regions. We denote by T_i the resultant tetracoordinated polytopes. The matching around the $\{3, 3, 5\}$ original edges involves

Table 3.3. *Comparison of inter-site distances between the Coleman–Thomas cluster and the {5, 3, 3} based cluster*

No. of neighbours	4	12	24	12	4	24	24	28
Coleman–Thomas	2.35	3.83	5.45	6.3	6.67	7.53	8.60	9.12
Polytope (on S^3)	2.35	3.82	5.45	6.27	6.62	7.43	8.48	9.08
Mapped polytope	2.35	3.83	5.47	6.3	6.66	7.49	8.56	9.17

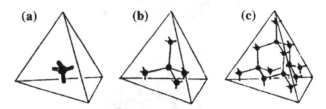

Fig. 3.3. First stages of decoration of the {3, 3, 5} tetrahedral cells. (a) 1 vertex, which generates the polytope {5, 3, 3}, (b) 5 vertices, (c) 14 vertices.

pentagonal rings. The first two stages of the decoration are shown on figure 3.3. Now it is possible to identify the large cluster built by Dandoloff *et al.* (1980) as mapped from T_1 and those studied by Gaskell (1975) as mapped from T_2. More precisely in the latter case, the 280-vertex model with icosahedral symmetry is $[T_2, C, \pi/5, d]$ while the 798-vertex model with tetrahedral symmetry is very close to $[T_2, V, 0.776rd, d]$ (figure 3.4).

3.2 Toroidal vesicles with phospholipid membranes

3.2.1 Toroidal topology and the Clifford torus

It is possible to obtain artificial vesicles with a phospholipidic membrane in water. The phospholipidic amphiphilic molecules form a bi-layer enclosing a finite volume of water. This example is very different from the above atomic clusters: this is a set of molecules, not in vacuum, but in water. Nevertheless, the contradiction between the structure of the fluid membrane, which prefers to be flat, and the high energy of its free border which tends to a closed structure, is a nice example of geometrical frustration.

Vesicles with toroidal topology are predicted to be stable for standard free energy models (Ou-Yang and Zhong-Can 1990, Seifert 1991, Michalet *et al.* 1994), and have been observed with partly polymerized phospholipid membranes (Mutz and Bensimon 1991) and more recently with fluid membranes (Michalet and Bensimon 1994). Furthermore, calculations and experiments

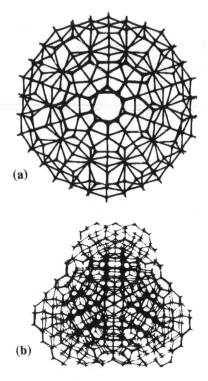

(a)

(b)

Fig. 3.4. Tetravalent clusters described by Gaskell (1975). (a) The 280-vertex model with icosahedral symmetry, (b) the 798-vertex model with tetrahedral symmetry.

agree on the specific toroidal shape, the so-called 'Clifford torus', an axial-symmetric torus such that the ratio of its generating circles is $\sqrt{2}$. From the theoretical side, this results from the consideration of the bending energy F_b which is proportional to the so-called 'Willmore functional' T (Willmore 1982, Pinkall and Sterling 1987):

$$T = \int H^2 \, \mathrm{d}A \qquad (3.2.1)$$

where H is the mean curvature. A second term in the energy comes from an integration over the Gaussian curvature which depends only upon the surface genus, and is therefore constant here. Willmore has conjectured that $T \geqslant 2\pi^2$ for any toroidal surface in R^3, the equality holding for the Clifford torus. A second important point is that the Willmore functional is invariant under a conformal transformation in R^3. This property was recalled by Duplantier (1990) in the context of vesicle geometry and extensively used by Seifert (1991) in deriving stability phase diagrams. Both authors note that this conformal transformation invariance implies fluctuations in the ground state configuration.

We are presently interested to show the following points (Mosseri *et al.* 1992).

- By lifting the vesicles into the spherical space S^3, embedded in R^4, the study of conformal transformation invariance is largely simplified. The set of toroidal shapes conformally equivalent, in R^3, to the Clifford torus is parametrized by a single angle ω in the range $[0, \pi/2]$.
- If an area (or volume) dependent term is added in the free energy, one expects configurational oscillations around the Clifford torus. For small enough α, it is analogous to a two-dimensional oscillator.

We shall see that the Clifford torus is a particular stereographic projection of the 'spherical torus' in S^3. This is very interesting in the context of the curved space approach to frustration in amphiphilic bi-layers. The central role played by the spherical torus as a frustration-free ideal membrane was stressed in the previous chapter. From this point of view, the Willmore functional gives indications for choosing the best mapping from the ideal surfaces in curved space to the real, Euclidean one. For instance, as one is interested in the way the bi-layer surfaces can increase their genus (even up to infinite genus configurations), the following theorem might prove useful (Willmore 1982): there exist surfaces of arbitrary genus in R^3 which are extrema of the Willmore functional; these surfaces are the stereographic projection of minimal surfaces in S^3. Note that vesicles of higher genus have recently been obtained (Michalet and Bensimon 1994). The present approach can be used to describe these vesicles and their conformal variants. To our present knowledge, there are no analytic coordinates for the higher genus minimal surfaces on S^3 and numerical work is therefore needed (Seifert 1993). These vesicles are nevertheless a very beautiful example of 'applied' topology, which have stimulated mathematicians. See, for instance, Kusner (1989), who has recently been interested in finding solutions to the Willmore problem.

3.2.2 *The torus foliation of S^3 and the Hopf map*

Points in the unit radius S^3 have the following coordinates x_i in a toroidal system (slightly different from that defined in appendix A1, §2.3):

$$x_0 = \cos\theta \sin\phi \quad x_1 = \sin\theta \sin\phi \quad \omega, \theta \in [0, 2\pi[$$

$$x_2 = \cos\omega \cos\phi \quad x_3 = \sin\omega \cos\phi \quad \phi \in \left[0, \frac{\pi}{2}\right]$$

The angle ϕ parametrizes a continuous foliation of tori with two exceptional circles ($\phi = 0$, $(\pi/2)$) as axes, three tori being displayed in figure A1.4, under a stereographic projection in R^3. Here the pole of projection is on one of the two exceptional circles. Consider the following Hopf map, also defined in appendix A3:

$$\psi: S^3 \to S^2 \quad (\theta, \phi, \omega) \to (\alpha, \eta)$$
$$\alpha = 2\phi, \quad \eta = \theta - \omega\lfloor 2\pi \rfloor, \quad \alpha \in [0, \pi], \quad \eta \in [0, 2\pi[\tag{3.2.2}$$

where α and η are standard polar coordinates on S^2. Note that here we do not use θ and ϕ for the S^2 angles, in order to avoid confusion with the same angles in S^3. This is just a manifestation that the Hopf bundle is not 'trivial', the base cannot be embedded in the fibred space!

This map is such that the inverse image of a point on S^2 is a great circle on S^3. The full pre-image of S^2 points is a fibration of S^3 by great circles, called the Hopf fibration (figures A3.2 and A3.3); S^2 is the 'base' of the fibration. Each torus of the above foliation maps onto a 'parallel' circle (constant α), the spherical torus ($\phi = (\pi/4)$) being mapped onto the equator.

3.2.3 Stereographic map and conformal transformations

Let us consider now the stereographic map of the spherical torus. By symmetry, all points of the same Hopf fibre, when used as a pole for this projection, give the same torus in R^3. Furthermore, the different shapes only depend on the distance, in the base space S^2, between the image of the spherical torus (the equator) and the image of the pole. Hence a given torus of the foliation is the locus of points in S^3 which, as the projection pole for the spherical torus, gives the same toroidal shape in R^3. The stereographic map, which is also a conformal transformation, is such that orthogonal transformations in S^3 correspond to inversions in R^3. Therefore, as far as the shape is concerned, changing the pole in S^3 will produce all the conformally equivalent objects in R^3. When the projection pole is on one of the two exceptional circles (north or south pole in the base), the spherical torus maps onto the Clifford torus in R^3.

As far as the shapes of the torus in R^3 are concerned (modulo rotations and dilatations), they are parametrized by the angle α in the range $[0, (\pi/2)]$. The calculation gives the following coordinates (X, Y, Z) for the torus in R^3 (recall $\alpha = 2\phi$):

$$X = \frac{\sin\phi\cos\omega - \cos\phi\cos\theta}{\sqrt{2} - W} \qquad Y = \frac{\sin\phi\sin\omega + \cos\phi\sin\theta}{\sqrt{2} - W}$$

$$Z = \frac{\sin\phi\sin\theta - \cos\phi\sin\omega}{\sqrt{2} - W} \quad \text{with} \quad W = \sin\phi\cos\theta - \cos\phi\cos\omega$$

(3.2.3)

For $\alpha = 0$, π, the Clifford torus is recovered. Figure 3.5 shows three conformally equivalent tori in R^3. With the aid of the 'Surface Evolver' program, Hsu *et al.* (1992) have produced very nice pictures of tori of high genus, which are reproduced in figure 3.6.

Fig. 3.5. Three stereographic projections in R^3 (conformally equivalent) of the spherical torus, obtained for three different positions of the projection pole.

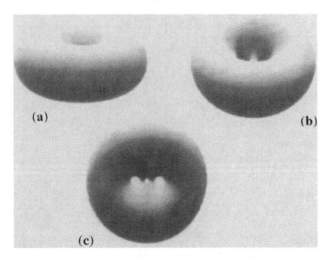

Fig. 3.6. Tori with genus 1 (a), 2 (b) and 3 (c) have been obtained from surfaces of the same topology, but having evolved in order to satisfy the Willmore minimization condition (eq. 3.2.1). They are very close to those experimentally observed. The torus of genus 1 is a Clifford torus (from Hsu *et al.* 1992).

3.2.4 Shape oscillations

Suppose that a term, a function of the area or even of the volume, is added to the energy. Since the area is minimal for the mapped Clifford torus ($\alpha = 0$), it remains as an equilibrium configuration (for some range of the volume to surface ratio). If a small conformal deformation is imposed, a force will oppose this deformation. This deformation is simply obtained by a displacement of the projection pole, characterized by an angle α. To first order (α small enough), the force is linear with the deformation, as in Hooke's law, and the physics is analogous to that of the harmonic oscillator. We have changed the problem of the shape oscillations into the problem of the oscillations of a point (the pole) on a surface (the base).

One expects a mode of shape pulsation, whose frequency is directly linked to the physical parameters. The symmetry of the problem is best seen by locating the projection pole onto the base space S^2. Once α is given, the angle η remains free, which shows that it is indeed like a two-dimensional oscillator. With increasing α, anharmonic effects are expected. Whether these modes can be experimentally measured is an open question. Note finally that we have only considered oscillations in a continuous set of conformally equivalent tori; other deformations, more complex, could also be possible (Fourcade 1992).

4

Decurving and disclinations

4.1 Disclinations

A disclination is a defect involving a rotation operation, as opposed to the more familiar dislocation, which is associated with a translation given by its Burgers vector (Friedel 1964). For this reason, this defect, introduced by Volterra at the beginning of the twentieth century in his description of a continuous solid medium, is sometimes called a rotation-dislocation. A disclination can be generated by a so-called 'Volterra' process, by cutting the structure along a line and adding (or removing) a sector of material between the two lips of the cut. In two dimensions, this defect is point-like, while it is linear in three dimensions. The two lips of the sector should be equivalent under a rotation belonging to the structure symmetry group in order to get a pure topological defect confined near the apex of the cut (Kléman 1983).

4.1.1 A simple example of disclinations: wedge disclinations in two dimensions

It is possible to describe this defect, and the induced deformation, as a concentration of curvature (figure 4.1). This will be argued, in §4.4, from a differential geometry analysis, but it is possible, in two dimensions, to describe this relation more simply. Let us first do the Volterra construction with a sheet of paper. We first cut it along a straight segment up to its centre. Then, upon rotating around this centre, we can either add or remove a sector, and then glue again along the lips of the cut. The angle of rotation is called the weight, or the angular deficit, of the disclination (even though the value can be positive as well as negative). As in figure 4.1, one gets a non-flat sheet of paper with either a conical or a saddle point singularity at its centre. Can we have a more quantitative measure of this curvature?

57

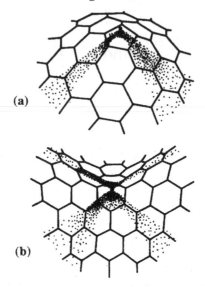

Fig. 4.1. Disclinations in a hexagonal structure. (a) Positive disclination. (b) Negative disclination.

Consider a non-flat surface. Its Gaussian curvature can be detected and measured by means of the parallel transport of a vector along a closed circuit. The simplest thing to do is to draw, on the surface, a polygonal line with geodesic edges (figure 4.2). Then, a vector is transported in such a way that its angle with the side along which it is transported remains constant. When, having travelled around the circuit, the vector returns to its starting point, it is found to have been rotated as compared to its initial orientation. This rotation is related to the integrated Gaussian curvature in the area enclosed by the circuit:

$$\delta = \iint_{\mathscr{D}} \kappa \, d\sigma \qquad (4.1.1)$$

where δ is the rotation angle of the transported vector. Analogously, on a surface punctured by disclinations, parallel transport of a vector along a circuit which encloses a disclination also results in a clear rotation of the vector: this rotation angle is exactly the angle of the disclination. Therefore, a disclination can be considered as a concentration of curvature.

4.1.2 Wedge disclination in three dimensions

A three-dimensional disclination can also be generated via a Volterra process by cutting the structure along a half plane and adding (or removing) a wedge of material between the two lips of the cut. The defect is now linear. The two faces of the wedge should be equivalent under a rotation belonging to the structure symmetry group in order to allow a perfect matching between the lips and the added wedge. A pure topological defect is then confined near the axis of the cut. In the case of wedge disclinations in three-dimensional space the axis of the rotation is located on the defective line (see figure 4.3).

It is then clear that wedge disclinations are located along geodesic lines of the structure. Indeed, if a disclination is introduced in a Euclidean space, the defective line is a straight line, which is invariant under the rotation defining the disclination. In a spherical three-dimensional space, the line invariant under a rotation is a great circle; if the sphere S^3 is embedded in R^4, this line is the intersection of S^3 with the two-dimensional plane invariant under the rotation. Then, in this case, the disclination line will also lie on a great circle. The fact that these defects are along geodesics can be shown analytically, using Bianchi identities (see §4.4).

4.1.3 Wedge disclination and curvature

As in two dimensions, wedge disclinations can be viewed as loci of curvature concentration in a three-dimensional space. If the disclination is obtained by a Volterra process in which matter has been removed (added), it is a positive (negative) disclination. The sign of a disclination can also be determined by parallel transport: if the transported vector rotates in the same (opposite)

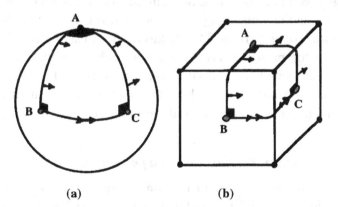

(a)   (b)

Fig. 4.2. (a) Parallel transport along a spherical triangle (A, B, C) having three $\pi/2$ angles. The transported vector rotates by $\pi/2$. (b) A cube is a topological sphere with concentration of curvature on its vertices. Parallel transport along a triangle (A, B, C) having three $\pi/2$ angles. Here again the transported vector rotates by $\pi/2$. Note that the cube is covered by eight such triangles, which amounts to a total curvature of 4π, which is precisely that of a (unit radius) sphere. In order to ensure continuity for the parallel transport in this case, the cube edge must be considered as a fold on a planar surface containing the two square faces sharing the edge.

direction as the circuit, the disclination is positive (negative). Therefore, introducing negative disclinations can be used in order to decurve a positively curved space.

Wedge disclinations are analysed in terms of curvature concentration in appendix A8 using differential geometry. Suppose that the curvature of a disclination is spread into a small tube surrounding it. The space is locally identical to $S^2 \times R$ (imagine some kind of generalized cylinder built with a sphere in place of the usual circle; such a space is presented in appendix A1). The Riemann tensor is given in equation (A8.23), assuming that the e_3 vector is along the disclination line. Any vector which is transported along a circuit enclosing the disclination line is rotated by the angular deficit δ. This is the limit of the equation (A8.21) used to define the coefficients in the Riemann tensor.

4.1.4 Effect of wedge disclinations

A two-dimensional structure can be described as a set of edges, joining vertices and separating polygonal cells. A disclination changes the coordination number when it goes through a vertex; it changes the number of sides of a polygonal cell threaded by the defect.

In three dimensions, disclinations also change the network topology. For example, as in two dimensions, they change the coordination number when they go through a vertex (figure 4.3). So, introduction of disclinations not only allows for decurving the embedding space, but it also generates slight modifications of the local configurations. This makes the disclination decurving mode very attractive in the present curved space approach, since it will introduce disorder in the ideal template, and will make it look closer to the amorphous structure. Furthermore, we shall see later that *intrinsic* disclination networks can also be found in some complex crystalline structures.

4.1.5 Disclinations in a polyhedron

We now show how to generate disclinations in the simple case of a polyhedron. We do not consider the solid (three-dimensional) object but the two-dimensional tiling on a topological sphere. The cut axis is aligned with one polyhedron symmetry axis and the Volterra construction is done. Figure 4.4 illustrates the case of a tetrahedron with disclinations along the three-fold and two-fold axes. The axes cut the polyhedral surface in two points, the defect being thus a couple of disclination points; the three-fold axis case changes the tetrahedron into a square pyramid. The two point-defects are a vertex (which

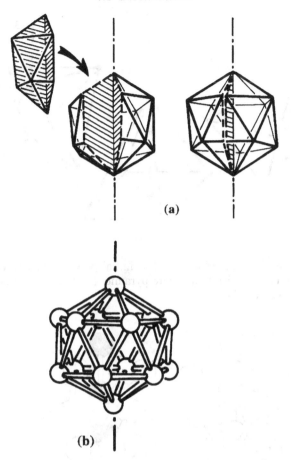

Fig. 4.3. (a) Procedure to insert a disclination. (b) Effect of a disclination on an icosahedral configuration: a Z_{14} coordination polyhedron has been generated.

becomes four-fold coordinated) and the centre of a triangle (which becomes a square). The two-fold axis case is here remarkable in that it transforms the tetrahedron into another regular (Platonic) polyhedron: the octahedron. In that case two opposite edges are transformed into regular triangles and all the vertices become four-fold coordinated.

4.1.6 Disclinations in a polytope

Consider a $2\pi/5$ disclination line in the $\{3, 3, 5\}$ polytope. The local order around such a line is shown in figure 4.3. In the $\{3, 3, 5\}$ polytope, the coordination polyhedron of a vertex is a regular icosahedron. The cut axis is aligned with a five-fold icosahedral axis and a $2\pi/5$ wedge is inserted. The

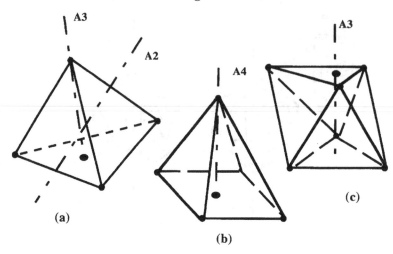

Fig. 4.4. Disclination in a tetrahedron. (a) The initial polyhedron. (b) The defect line along a three-fold axis gives rise to a square pyramid. (c) The defect line along a two-fold axis gives rise to an octahedron.

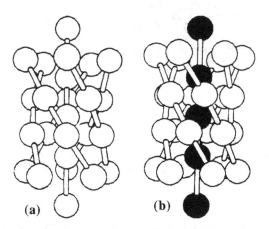

Fig. 4.5. Local configuration of sites in S^3. Only intra-fibre edges are drawn. (a) In the $\{3, 3, 5\}$ polytope each arrangement of neighbouring sites is surrounded by five similar sets. Note that the mapping has transformed the central great circle sub-set (which contains the projection pole) into a straight line and the other circles into helices. (b) When a disclination is introduced, the line defect (with sites in black) is now surrounded by six similar sets.

new coordination polyhedron is a 14-vertex triangulated structure and the central vertex is called a Z_{14} site in the standard notation due to Frank and Kasper (see appendix A6). The central site and the two opposite vertices on the cut axis belong to a $2\pi/5$ disclination line. It is possible to show (see §4.4) that such a line cannot stop in the material and either ends at the surface, splits at crossing points or forms closed lines. The set of crossing points and disclination segments or lines forms a so-called disclination network. This kind of network, which plays an important role in the description of complex structures, will be investigated in the next chapter.

4.1.7 A pair of disclination lines in polytope {3, 3, 5}

We first describe more precisely the generation of two defect lines in the {3, 3, 5} polytope. Let us consider the 'spherical' torus defined by $\phi = \pi/4$ in the S^3 torus foliation (see appendix A1). It can be approximated by a 5×5 piece of square tiling with opposite sites connected. It leads to a partition of S^3 by ten pentagonal prisms which can be filled with {3, 3, 5} vertices (see figure A5.2). Here the {3, 3, 5} polytope is oriented in such a way that two five-fold axes coincide with the two great circles, principal axes of the foliation, defined by $\phi = 0$ and $\phi = \pi/2$. Each pentagonal prism can still be cut into five trigonal prisms. A $2\pi/5$ disclination line can be generated by inserting a sixth such trigonal prism. This can be done simultaneously for the two sets of five pentagonal prisms, leading to two sets of hexagonal prisms. Only the sites located on the great circles $\theta = 0$ and $\pi/2$ have their local environment modified; they become Z_{14} sites. The new polytope contains 168 sites, 144 twelve-fold coordinated vertices (Z_{12} sites) and 24 Z_{14} sites. The combination of two disclinations along opposite great circles can be called a screw disclination since it involves simultaneous rotations around two completely orthogonal planes; however the translational part of a usual screw is replaced here by a second rotation in S^3.

It is here interesting to recall the work of Straley (1985), who studied, by numerical simulation, the ground state of 168 spheres on S^3, under a standard pair potential, and has obtained this same polytope, with two disclinations.

4.1.8 More disclinations in the {3, 3, 5} polytope

We now briefly describe a method which generalizes the previous approach to the case of several tangled defect lines (Nicolis *et al.* 1986). Indeed the two above great circles are members of a Hopf fibration (see appendix A3). These two circles are symmetry axes of the polytope. The fibration contains other symmetry axes of equal order. These axes can be distinguished using the Hopf map. This map is such that the inverse image of a point on S^2 is a great circle on S^3. The full inverse image of S^3 gives a fibration of S^3 with great circles, called the Hopf fibration. Each fibre is tangled with any other fibre. Now if a polytope symmetry axis is a member of the fibration, other axes, conjugate to the previous one, also belong the fibration.

In the case of the five-fold axis, the great circle contains 10 vertices (it is also a 10-fold screw axis). There are 11 other axes belonging to the fibration, each polytope vertex being on exactly one such axis. The Hopf map of these 12 axes leads to 12 points on the base S^2 which form a perfect icosahedron. The symmetry on the base S^2 is reminiscent of that on S^3. Consider one point on S^2 and its five neighbours. The pre-image on S^3 is a great circle (containing 10 polytope vertices) surrounded by five similar great circles as shown in figure 4.5a. Disclinations on S^3 can be introduced by first disclinating on S^2 and then inverting the Hopf map. In the case considered in figure 4.5b, two $2\pi/5$ disclination points are created on S^2 which leads to two $2\pi/5$ disclination lines on S^3. One recovers the above discussion leading to polytope '168'. The local order along one such disclination in S^3 is shown in figure 4.5. The line defect (with dark sites) is now surrounded by six lines of the fibration.

It is further possible to generate three or four defect points on S^2, leading to three or four defect lines and the polytopes '180' and '192' according to their number of vertices. These two polytopes have also been called D_{15} and D_{16}. This is to recall that they give, under a Hopf map, a Z_{15} and a Z_{16} polyhedron on S^2. For the same reason, polytope '168' is also called D_{14}.

The radial distribution function of these polytopes is shown in figure 4.6 and compared to the original $\{3, 3, 5\}$ radial distribution function. New distances have appeared which can be interpreted as a broadening of the $\{3, 3, 5\}$ distance distribution. Scaled to the first neighbour distance, the radius of S^3 has increased from the $\{3, 3, 5\}$ to the disclinated polytope values, which shows that it is indeed a decurving process.

4.1.9 Disclination in the {5, 3, 3} polytope

To each disclinated $\{3, 3, 5\}$ polytope, there corresponds simply a disclinated $\{5, 3, 3\}$ obtained upon dualizations. For example the polytope '864' with a pair of $2\pi/5$ disclinations has been previously described and its excitation spectrum calculated (Mosseri *et al.* 1985). In this case the vertices remain tetra-coordinated, but the dodecahedral cages threaded by the disclination lines are changed into polyhedra with twelve pentagonal faces and two hexagonal faces.

4.2 Wedge and screw disclinations

The Burgers vector of a dislocation can be oriented in different directions relative to the dislocation line. These two basic cases are edge dislocations and

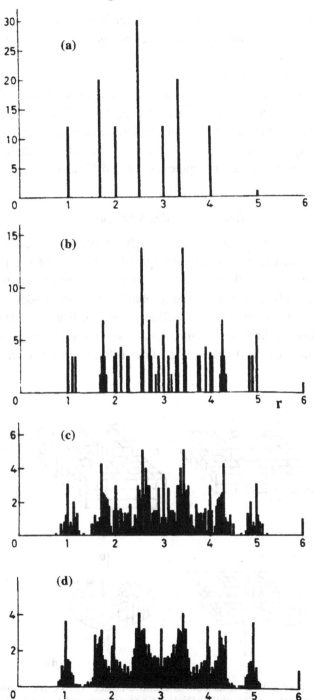

Fig. 4.6. Radial distribution function for disclinated polytopes. (a) $\{3, 3, 5\}$ Polytope. (b) Polytope D_{14} with two disclination lines. (c) Polytope D_{15} with three disclination lines. (d) Polytope D_{16} with four disclination lines.

twist dislocations. Harris (1977, 1978) has described the same possibilities for disclinations. The axis of rotation can be either parallel or orthogonal to the defect line.

Figure 4.7 presents some examples of rotation and translation defects. Note that one can also consider combined defects, called 'dispirations'. If we restrict the observation of these defects in a finite region excluding the core of the defect, a large number of combined defects can be defined. But some difficulties appear at long distances from the defective line. The example of a screw disclination is helpful in order to appreciate these difficulties.

4.2.1 Screw disclinations

A screw disclination can be generated through a Volterra process: the two lips of the cut are rotated along an axis orthogonal to the cut before gluing. This rotation axis is therefore orthogonal to the disclination line which is the border of the cut. The defect line must be globally invariant under the rotation involved in the Volterra process. In Euclidean space it is a circle and therefore not a geodesic line of the space. Figure 4.8 shows how such a defect appears in a cubic lattice. In Euclidean space, it is rather difficult to follow the trajectory

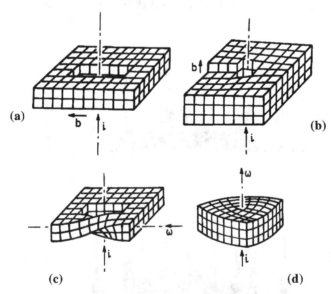

Fig. 4.7. Local representation of defects in a cubic lattice. (a) Perfect lattice. **i** defines the line where to insert the defect. (b) Dislocations correspond to a relative translation of the two lips of the cut: here a screw dislocation with Burgers vector **b**. (c) and (d) Disclinations correspond to a relative rotation of the two lips. Rotations are defined by the ω vector. A screw disclination (c) and a wedge disclination (d) are generated.

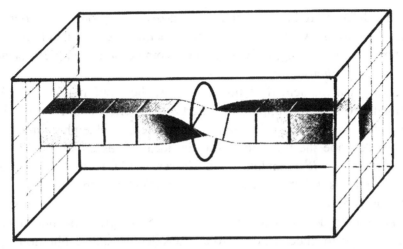

Fig. 4.8. A disclination line forming a loop. This loop can be obtained by a Volterra process: the cut is defined in the plane outside the loop. Then a lip of this cut is rotated using a rotation axis normal to the loop in its centre; this process lets the circular loop be invariant. Under gluing the two lips, a screw disclination is obtained. (After W. Harris.)

of all the points of the cutting half plane during the Volterra process. This difficulty mainly arises from the infinite extension of the Euclidean space. So, at least formally and to some extent, screw disclinations are easier to describe in a three-dimensional spherical space.

Consider S^3 dressed with a Hopf fibration (appendix A3). The screw disclination line is introduced on a great circle of the fibration (for example the one which maps onto the south pole on the base S^2) and the rotation axis of the Volterra process lies on a completely orthogonal fibre (whose Hopf map then gives the north pole on the base). The Volterra cutting surface is in this case a half great sphere (S^2), embedded in S^3, which is limited by the disclination line, and pierced orthogonally by the rotation axis. In this case, the disclination line is a geodesic line, which is globally invariant under the rotation, whatever the rotation angle is. So it is possible to glue the two lips of the cut when one is rotated relative to the other with this rotation axis. Since the two lips of the cut are rotated relative to an axis orthogonal to their surface, there is no need to remove (or to add) in order to perform the Volterra process: *the screw disclination does not carry curvature.* The effect of the Volterra process on the spherical torus (which is the surface equidistant from the two particular fibres used to define the defect) is nevertheless significant. The torus is cut along a small circle, then twisted and re-glued. If we follow a fibre of the Hopf fibration belonging to the torus, it is cut and reconnected to another fibre on the torus.

Prior to the defect introduction, each fibre makes exactly one turn around the two axes of the torus. After the Volterra construction, the modified fibres make several turns before closing, if the rotation is commensurate with 2π, or never close if it is incommensurate.

In fact, from a mathematical point of view, the Hopf fibration has been transformed into a Seifert fibration (Seifert 1933, Seifert and Threfall 1980). Such an object has already been used, in the context of blue phase modelling, in order to describe the transition from the ideal configuration on S^3 into a double-twist tube in R^3 (Dandoloff and Mosseri 1987). Note that every other torus on the torus foliation (with the same axes) presents the same new topology for its fibres.

Inserting such a defect in the discrete $\{3, 3, 5\}$ polytope is another useful exercise in order to understand its role. Consider the above fibration of the $\{3, 3, 5\}$ polytope by twelve circular fibres, each gathering ten vertices. These circles are five-fold axes. The disclination is located on the fibre corresponding to the south pole of the base, and the rotation axis which builds the defect is on the completely orthogonal fibre (north pole on the base). Let us consider a $2\pi/5$ rotation. This introduces a chirality in the structure but does not change the number of vertices. Here we point again to a main difference between wedge and screw disclinations, since, in the former case, matter is added (or subtracted) while, in the latter, the matter content is constant. The disclination effect can be followed on the two axes and on the two tori on which the 120 vertices are initially located: ten vertices on each axis and fifty vertices (five fibres) on each torus (see appendix A5, §2).

- The rotation axis (north pole on the base) does not change.
- The first torus is cut along a small circle (a pentagon of five edges) and then reglued after a $2\pi/5$ rotation. As a consequence the five fibres are now interconnected, leading to a single screw curve making four turns around one (north) torus axis, and five turns around the other one (or six and five turns for the fibres parallel to the second diagonal; see figure 4.9). Note that this transformation keeps unchanged the local topology of the vertices lying on the first axis, as well as those lying on this torus: they keep an icosahedral coordination first neighbour shell, with only small metrical distortions.
- If we consider the second torus as a cylinder with its two ends identified, this torus is cut along a generatrix of this cylinder. Then, one lip is rotated relative to the other (if the torus is represented by a cylinder this appears as a translation, or a shear, along a straight generatrix). Here again, the five fibres are changed into a single curve making the same number of turns around the two axes as for the first torus.
- The behaviour of the ten vertices of the last axis, on which the defect is located, is

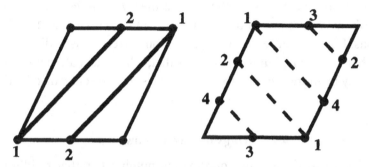

Fig. 4.9. How a screw disclination changes the spherical torus, with its Hopf fibres. In this example, the screw disclination, located along an axis of the torus, is a π disclination. The torus, seen as a square with identified opposite sides, is distorted into a rhombus. Indeed, if one considers a side of the square as a lip of the Volterra cut, then, once unfolded, a rotation by π of the lip corresponds to a translation by a half edge length. Opposite sides of the rhombus should be identified. Two fibres f_1 and f_2, which are parallel to a diagonal of the square, lead to a single line. This line makes one turn around a torus axis, and two turns around the other axis. But, if we consider another family of fibres parallel to the second diagonal of the square (f_1' and f_2'), after the disclination procedure, we observe that the resulting line makes three turns around a torus axis, and two turns around the other axis. So the winding number depends on the relative orientation of the twist angle of the fibration and the disclination angle.

much more complex because there is no simple way to localize the vertices after the defect has been inserted. Suppose that in place of these vertices we have small balls. These balls are threaded by the defect line which is the border of the Volterra cutting surface. The two lips are displaced by a large quantity (approximately two edge lengths). The balls experience a large stretching along the line and their precise location is therefore lost in the core of the defect. A reasonable solution is to take this fibre unchanged with ten regularly disposed vertices, but there remains in fact some arbitrariness for the position of these vertices. It is impossible to maintain the local $\{3, 3, 5\}$ polytope connectivity (the icosahedral environment) since some edges are strongly elongated. New local configurations appear for the vertices on the defect line and on the surrounding torus, with coordination numbers 13 or 11. This can be described by saying that the screw disclination is dressed by an interwoven network of positive and negative wedge disclinations. Indeed, as said above, screw disclinations do not carry curvature. Consequently the equivalent interwoven network of wedge disclinations must contain positive as well as negative disclinations in such a way that these opposite defects balance each other. Related problems are discussed below in the section devoted to disclination density and Regge calculus, and in appendix A6, §2.

Two geometrical quantities are often invoked in the characterization of a manifold: the curvature and the torsion. See for instance Rivier's (1990) point of view of differential geometry applied to the theory of defects in a continuous medium. We have already said that the curvature is associated with wedge disclinations. It is sometimes argued that torsion corresponds to dislocations (de Wit 1981, Kröner 1981); but it seems to us that, when translational symmetry is lacking, it may be more accurate to associate torsion and screw disclinations.

4.2.2 Rings in covalent systems

In covalent systems, the atomic coordination number is rather rigidly fixed by the strong chemical bond, which in addition constrains the range of bond angle variation. When disorder dominates, it manifests itself usually at the level of ring configuration. At the descriptive (topological) level, there is a main distinction between two types of rings: those which are faces of cage-like local structures, and those which do not have this property. The $\{5, 3, 3\}$ polytope is for instance a model in which all minimal rings (pentagons) are faces of cages (dodecahedra). The polytope '240', and the diamond structure, are models in which all minimal rings (non-flat hexagons) cannot be assigned as faces of cages. In both examples two adjacent edges belong to several rings (more than two in a three-dimensional model).

Screw disclinations are rather special here since they can change rings which define cage faces into rings which do not define faces. A screw disclination in a hypercube $\{4, 3, 3\}$ provides such an example. A $\pi/2$ screw disclination located along a four-fold axis in S^3, which coincides with the symmetry axis of squares, transforms these squares into skew 5-gons, as described in figure 4.10. All vertices remain tetracoordinated. Among the eight cubes of the hypercube, those which are not threaded by the disclination line are topologically unchanged, while the remaining four cubes disappear.

An even simpler insight into this kind of defect can be obtained by considering a screw dislocation in a linear arrangement of cubes in three dimensions. This structure, which is made of cages (the cubes), has two kinds of squares: those lying on the surface, and those pierced by the four-fold axis. Let us

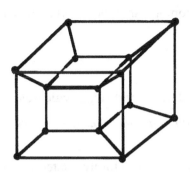 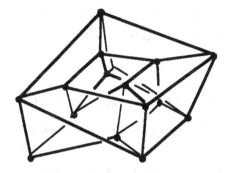

Fig. 4.10. A hypercube and the new structure which results from a screw disclination applied along a four-fold axis with the rotation along the opposite axis. In this schematic picture of the hypercube, one should imagine the rotation axis as a loop threading the four squares which connect the large external and the small internal cube (recall that all these cubes have equal size in R^4). Now, in this representation, the disclination line is along the vertical four-fold axis. Upon disclinating, some cubic cells (like the cell represented by the small internal cube, and the cell represented by the outside of the large cube) disappear. Only a toroidal cell remains.

introduce a screw dislocation along this axis, whose Burgers vector modulus equals the square edge length. The dislocation leaves the squares of the first kind unchanged, but changes the other ones into pentagons by adding a new edge, parallel to the Burgers vector, in each such square. This edge, and a congruent one, now define two rings: a pentagon and an unchanged square. The new, dislocated, structure does not contain polyhedral cells anymore.

4.2.3 Disclination in the polytope '240'

It is possible to introduce a pair of disclinations along two opposite 30/11 axes in this polytope. Since the defect lines pass through the middle of the channels, they avoid the sites which remain four-fold coordinated. But in this case the defects are related to a symmetry which is not a rotation: a twisted disclination is generated. We have generated two such new polytopes, with 384 and 360 vertices, respectively (Mosseri and Gaspard 1987). The latter is very interesting in that the two defect lines thread seven-fold rings, while all the remaining rings are even-membered. These lines are therefore instances of Rivier lines (Rivier 1979) whose effect on the excitation spectrum has been investigated (Baskaran 1986). Even more, the '240' polytope, which is related to the Connell–Temkin model (Connell and Temkin 1974), is a good model for amorphous III-V alloys, where the chemical order favours even rings with respect to the alternation of the two kinds of atoms. This '360' polytope is a good model to study the effect of 'false' homopolar bonds occurring with stoichiometric defects (see §4.5.4).

4.3 Coordination number, disclination density and Regge calculus

4.3.1 Two-dimensional systems: Euler–Poincaré and Gauss–Bonnet relations

Continuous and discrete two-dimensional systems have closely related properties, owing to the existence of two famous relations: the Euler–Poincaré and Gauss–Bonnet relations. The former is of purely topological nature, while the latter relates the topology and the metrical properties of the underlying space, through its Gaussian curvature. We shall see how various interesting results can be derived, which constrain any tiling in a two-dimensional manifold.

It is well known for instance that in the Euclidean plane the average coordination number of a triangulated structure (defined for example by the Dirichlet–Voronoi construction) is exactly six. This is immediately satisfied by the regular triangular tiling, in which case here all the vertices are six-fold coordinated, but it remains true for any disordered triangulated tiling. This type of result can be generalized easily to two-dimensional curved spaces (Riemann manifold), in which the metrics determine the tiling statistics.

Let us consider a regular two-dimensional manifold and let κ be its Gaussian curvature. In a finite domain, $\mathscr{D}$, not necessarily simply connected (it can contain holes), let F be the total number of faces, F_p the number of faces with p edges, E the number of edges and V the total number of vertices, including those on the frontier Γ of the domain $\mathscr{D}$. Let us restrict the calculation to the case where the connectivity is a constant c for internal vertices and c_r ($r = 1 \ldots n$) on the frontier; n is the number of vertices on the frontier (figure 4.11).

The calculation follows two steps: first, the Euler–Poincaré relation assigns a value to the alternate sum $V - E + F$, which is a topological invariant, called the Euler–Poincaré characteristic χ:

$$V - E + F = \chi = 2 - 2g \tag{4.3.1}$$

where g is the genus (number of holes) of the surface. If the domain is locally homomorphic to a bounded domain of a plane, then $\chi = 1$. The Gauss–Bonnet relation is then applied to take into account the metrical properties of the underlying space.

The total number of faces is obviously:

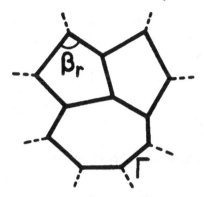

Fig. 4.11. Schematic drawing showing the internal angle β_r and the frontier Γ of a two-dimensional domain $\mathscr{D}$ with $n = 11$, $V = 12$, $E = 14$, $F = 3$, $c = 3$ and $c_r = 2$ or 3.

$$F = \sum_{p=3}^{\infty} F_p \tag{4.3.2}$$

and the number of edges reads

$$E = 1/2 \sum_{p=3}^{\infty} pF_p + n/2 \tag{4.3.3}$$

The number of edges is also related to the coordination number

$$E = c(V - n)/2 + \sum_{r=1}^{n} c_r/2 \tag{4.3.4}$$

After some trivial algebra, one obtains from equations (4.3.1) to (4.3.4) the first important expression

$$\sum_{p=3}^{\infty} (2c - cp + 2p)F_p = 2c\chi - n(c + 2) + 2 \sum_{r=1}^{n} c_r \tag{4.3.5}$$

Now let us focus on the metrical properties of the tiling. The Gauss–Bonnet relation reads

$$2\pi\chi + \sum_{r=1}^{n} \beta_r - n\pi - \iint_{\mathscr{D}} \kappa \, d\sigma = 0 \tag{4.3.6}$$

β_r is the internal vertex angle at the rth vertex on the boundary. We make the simplifying assumption that, at any vertex, the angles are all equal to $2\pi/c$, so that the internal angle at a vertex of the frontier Γ is

$$\beta_r = 2\pi(c_r - 1)/c \tag{4.3.7}$$

The edges of the polygons are also assumed to be along geodesic lines. Note that this regularization procedure can always been performed on smooth surfaces by small variations.

Then, considering equations (4.3.6) and (4.3.7) leads to

$$\sum_{p=3}^{\infty}(2c - cp + 2p)F_p = \frac{c}{\pi}\iint_{\mathscr{D}} \kappa \, d\sigma \qquad (4.3.8)$$

In particular, for coordination number three, equation (4.3.8) becomes

$$3F_3 + 2F_4 + F_5 - F_7 - 2F_8 - 3F_9 - \cdots = \frac{3}{\pi}\iint_{\mathscr{D}} \kappa \, d\sigma \qquad (4.3.9)$$

The hexagonal tiling of the plane corresponds to $0 \times F_6 = 0$ since the Gaussian curvature of the plane is zero. For all the polyhedra with coordination c, as $\iint_{\mathscr{D}} \kappa \, d\sigma = 4\pi$, it follows that the right-hand side of equation (4.3.8) is $4c$. In particular if $c = 3$ the right-hand side of equation (4.3.9) is 12. This is easily verified for the tetrahedron ($3F_3 = 12$), the cube ($2F_4 = 12$) and the dodecahedron ($F_5 = 12$).

For coordination number $c = 5$ equation (4.3.9) gives

$$F_3 - 2F_4 - 5F_5 - 8F_6 \cdots = \frac{5}{\pi}\iint_{\mathscr{D}} \kappa \, d\sigma \qquad (4.3.10)$$

Applying the latter equation to a spherical polyhedron, we recover the icosahedron case with its 20 triangular faces. Coordination numbers of four and six give analogous results, given explicitly in equations (4.3.12) below.

It is worth noticing that the term in parentheses in equation (4.3.8) is symmetric in c and p. This is easily understood with the Voronoi–Dirichlet dualization scheme, where vertices are permuted with cells: as a result, the vertex coordination number translates to the size of the polygonal cells.

Note that polygons for which

$$(2c - cp + 2p) = 0 \qquad (4.3.11)$$

do not contribute to the left-hand side of equation (4.3.8) (the topological side): they act as neutral charges. From a geometrical point of view, one sees from equation (4.3.7) that their underlying geometry is flat. First values of c and p for the neutral charges are

$$c = 3, 4, 5, 6, 7$$

$$p = 6, 4, 10/3, 3, 14/5$$

The presence of fractional numbers means that, in that case, one needs at least a polygon pair to realize a neutral topological charge. The total number of neutral charges is undetermined, these being either isolated (hexagon with $c = 3$) or dipoles (a heptagon with a pentagon again in the case if $c = 3$). When the polygons are considered isolated, the charge associated with an elementary polygon is the excess (or deficit) in its number of edges relatively to the neutral polygon.

More precisely, the following sum rules hold for different connectivities. Let us call $2c\chi'$ the right-hand side of equation (4.3.8). χ' is equal to χ for an unbounded surface. If the surface has a frontier it is deduced from equation (4.3.6) and inherits a boundary contribution. Then

$$3F_3 + 2F_4 + F_5 - F_7 - 2F_8 - 3F_9 = 6\chi'c = 3$$

$$F_3 - F_5 - 2F_6 - 3F_7 \cdots = 2\chi'c = 4$$

$$F_3 - 2F_4 - 5F_5 - 8F_6 \cdots = 10\chi'c = 5$$

$$F_4 - 2F_5 - 3F_6 - \cdots = 3\chi'c = 6$$

(4.3.12)

Equations (4.3.12) allow us to obtain simple parity sum rules (Gaspard *et al.* 1984). Using the fact that $\sum_{p=3}^{\infty} F_p$ has the same parity as $\sum_{p=3}^{\infty} n_p F_p$ when n_p are odd, we deduce the following rules after subtracting the even terms on the left-hand side of equation (4.3.8). For $c = 3$, 4 and 5, the number of odd faces is even, while for $c = 6$ the number of even faces depends on the parity of χ'. If χ' is even (odd), the number of even faces is even (odd). In fact, one observes for $c = 3$, 4 and 6 that the parity conservation rule is directly linked to the evenness (oddness) of the neutral charge. (For $c = 5$ the neutral charge corresponds to a non-integral value $10/3$.) If 'the neutral charge' is an even polygon, the number of odd faces is even: conversely if it is an odd polygon, the number of even faces has the same parity.

The above parity rules generalize a result of Rivier (1979). For $c = 6$, no positive quantity enters the left-hand side of equation (4.3.8): $\chi' = 0$ then corresponds to the plane or the torus which can be tiled with triangles. But it is impossible to incorporate larger polygons because they cannot be compensated by charge of opposite sign (we exclude digons), still keeping the connectivity equal to six everywhere. In addition it is impossible to tile the sphere ($\chi' = \chi = 2$ or more generally any surface with positive Euler–Poincaré characteristics) with an average coordination of six. The parity rule, deduced from equation (4.3.12) with $c = 6$, will prove interesting when applied to surfaces with negative χ': tubular structures, like periodic minimal surfaces, or the hyperbolic plane. Such surfaces are used in chapter 7 in order to describe crystals of films built by amphiphilic molecules. We also briefly introduce minimal surfaces in the next paragraph in the context of generalized fullerene structures.

By reference to a perfect type of polygon (hexagon, for instance), we can consider that other polygons have a disclination point at their centre. Recall that we can have the same statement for vertex coordination instead of polygon size (symmetry between c and p). Halperin and Nelson (1978) used such a microscopic geometrical description in their approach to two-dimensional melting. When the temperature increases in a dense regular triangular packing, the first kinds of defect which appear are dipoles of dislocation, or 'quadri-disclinations', as a dislocation is a pair of disclinations. Above the 'hexatic' transition dislocation dipoles are dissociated into single dislocations. A simple example of a dipole of positive and negative disclinations is represented for the hexagonal tiling of the plane in figure 4.12. This dipole is a dislocation, as can be verified using a Burgers contour. Owing to the presence of these disloca-tions, the hexatic phase loses translational order, but keeps an orientational order, which eventually disappears at higher temperature when the dislocations further dissociate into a 'plasma' of isolated disclinations, a state associated with the two-dimensional melting. A similar approach was applied later to model melting on the hyperbolic plane (Nelson *et al.* 1982).

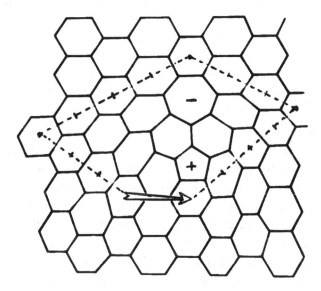

Fig. 4.12. A positive (pentagon) and a negative (heptagon) disclination in a hexagonal tiling of the plane. This dipole acts as a dislocation, as shown using a Burgers contour.

4.3.2 A physical example of a two-dimensional polygonal surface: the C_{60} fullerene

Structures built on three-dimensional curved spaces cannot model real structures, for the obvious reason that S^3 cannot be embedded in R^3. But this is not true for two-dimensional curved space, since S^2 can be embedded in R^3, and it is therefore interesting to look for the realization of two-dimensional spherical structures in Nature. The C_{60} fullerene, which was recently synthesized, is a fascinating example of such a curved structure, which illustrates applied topology at an atomic level.

The hexagonal graphite has well defined planes tiled by hexagonal 'neutral charges' with coordination $c = 3$, associated with carbon sp^2 hybridized orbitals. But it seems that this hybridization is not rigidly imposed by the electronic structure. Recall that the diamond form of carbon has four-fold coordinated atoms with sp^3 types of bonds, and even that small carbon clusters adopt linear forms with the associated sp orbitals. So, defects (with respect to pure graphite structure) are expected to occur, in the form of polygons whose internal angles are not too far from $2\pi/3$. The best candidates are therefore five-fold and seven-fold rings, the former being more probable since the internal angle is quite close to that of sp^3 hybrids. In that case, the underlying geometry inherits positive curvature and equation (4.3.9) imposes, if the surface becomes closed (topologically equivalent to a sphere), that there are exactly 12 pentagons. More precisely, if heptagons are allowed, the constraint is that the number of pentagons minus the number of heptagons gives 12. The hexagons are the neutral charge of the problem and their number is free. It seems however that

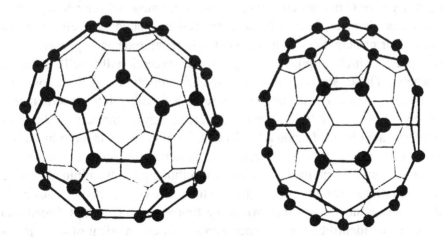

Fig. 4.13. A C_{60} molecule having a 'soccer ball' shape and a C_{70} molecule with more hexagons.

there is a net tendency to favour a very symmetrical shape with 12 pentagons and 30 hexagons, the so-called 'soccer ball' icosahedral polyhedron, with 60 carbon atoms (see figure 4.13). But several other molecules with more hexagons have already been experimentally observed, showing once more that Nature is an expert in geometry, since they all satisfy the above topological constraints (but had she any choice?). The simplest is the C_{70} molecule: using the Euler relation (4.3.4) it follows that $V = 70$ $F = 37$ and $E = 105$, where V, E and F are the numbers of vertices (carbons), edges (bonds) and faces (12 pentagons and 25 hexagons) (Fowler 1991).

Larger molecules, which keep the icosahedral symmetry, have been found or conjectured, with 240, 540, 960 ... atoms. These polyhedra are the dual of the Buckminster Fuller geodesic domes.

Other geometries have also been observed like those tubules which can be described as rolled graphitic planes closed at their ends by hemispherical cups, cut in C_{60} molecules (Ajayan *et al.* 1993). A one-dimensional metallic behaviour is expected in some cases (Mintmire *et al.* 1994).

More conjectural is negatively curved graphite (Mackay and Terrones 1991, Lenosky *et al.* 1992). In this case heptagons or even octagons come into the picture and introduce therefore some negative curvature. A hyperbolic infinite molecule could provide a possible solution, but here again only as an ideal model since the hyperbolic plane cannot be embedded in the three-dimensional Euclidean space. Notice that similar models, constructed on a hyperbolic plane, have been proposed for covalent tetracoordinated amorphous structures (Kléman and Donnadieu 1985) or for other kinds of glasses (Furtado and Moraes 1994). Another possible solution, embeddable in R^3, but with a less homogeneous negative Gaussian curvature, consists of using infinite periodic minimal surfaces. Such surfaces will be described in chapter 7. Being minimal surfaces, their principal curvatures are equal in magnitude and opposed in sign at any point, which imposes a negative (or exceptionally null) curvature everywhere. These surfaces can be tiled, for instance, by hexagons and heptagons (or octagons), leading therefore to a possible infinite structure of negatively curved graphite (with a reasonable energy). Infinite periodic minimal surfaces are obtained using a fundamental region which is reproduced by a translation group, and which also allows us to define the genus of the surface. Indeed, one would be inclined to assign to all infinite periodic minimal surfaces the same infinite genus. In order to differentiate, one considers the genus of the closed compact surface obtained when the fundamental region is folded onto itself upon side identification. In other words, it is the quotient of the minimal surface symmetry group by its translation group. The simplest infinite minimal surfaces in R^3 have genus $g = 3$.

As a torus of genus $g = 1$ can be seen as a square whose opposite sides are identified, similarly a torus of genus $g = 3$ can be seen as a 12-gon whose opposite sides are suitably identified. More generally a $4g$-gon leads to a torus with genus g.

Equation (4.3.8) can be applied to these compact surfaces (without border) with $c = p = 4g$. It follows:

$$\pi(4 - c) = \iint \kappa \, d\sigma$$
(4.3.13)
$$4\pi(1 - g) = \iint \kappa \, d\sigma$$

In the case $g = 3$ we have $\iint \kappa \, d\sigma = -8\pi$.

Consider now such a surface tiled by hexagons and octagons, with three-fold coordinated vertices. Equation (4.3.9) gives, together with equation (4.3.13):

$$F_8 = 12$$
(4.3.14)

The number of hexagons F_6 is still undetermined, the hexagon being the neutral charge. Figure 4.14 shows how to decorate a hexagon of $\{6, 4\}$ tiling of an infinite periodic minimal surface in order to have a three-fold coordinated tiling by hexagons and octagons. Octagons appear around four-fold coordinated

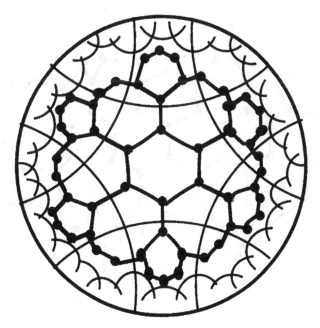

Fig. 4.14. Decoration of a hexagon of the $\{6, 4\}$ hyperbolic tiling, which introduces octagons around vertices.

vertices of the initial tiling. There are 6/4 octagons in each hexagon of the {6, 4} tiling. As there are eight hexagons in the fundamental region we still get as expected $F_8 = 12$. See also figure 4.15.

4.3.3 Coordination number in three-dimensional sphere packings

We have just seen above how the analysis of two-dimensional tilings, embedded on any surface, is greatly simplified thanks to the Euler–Poincaré and Gauss–Bonnet relations. We now try to obtain similar information about three-dimensional packings. However, if it is still possible to find extensions of topological relations like the Euler–Poincaré relation in higher dimension, the metrical part misses a relation like the two-dimensional Gauss–Bonnet relation. In addition, since the new dimension adds a new parameter, the number of cells, the topological relation proves itself less rich than in two dimensions. It is nevertheless possible to get useful – but non-rigorous – results by doing some approximations. Let us begin by looking at the determination of an average coordination number in sphere packings (recall that this coordination equals six for any disc tiling in two dimensions). The coordination number of a

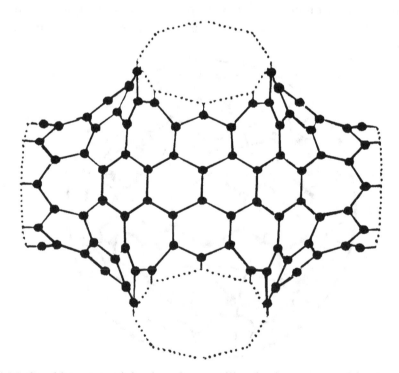

Fig. 4.15. Graphite on a minimal surface: a tiling by hexagons and heptagons (see Mackay and Terrones 1991, Lenosky *et al.* 1992).

site is equal to the number of faces of its associated Voronoi polyhedron. The packing units are supposed to be tetrahedra, which is quite natural for a disordered sphere packing, but can also be realized for any point set in space using a generically non-ambiguous simple decomposition. Voronoi vertices are therefore tetravalent. Each such vertex belongs to four Voronoi polyhedra and has three neighbours on each polyhedron. Any ring is common to two Voronoi polyhedra associated with two neighbouring sites. The ring size is equal to the number of tetrahedra sharing the two sites (sharing the edge joining the two sites in the original simple set).

If the tetrahedra were all regular, there would be room for about 5.1 tetrahedra around a common edge. Imagine now an 'impossible' structure, inside which each edge is shared by exactly $\bar{p} = 2\pi/\cos^{-1}(1/3) \simeq 5.104\,299$ tetrahedra. Such a structure has been proposed by Coxeter (1958) under the name 'statistical honeycomb', denoted $\{3, 3, \bar{p}\}$ (see appendix A4, §2.2). In the present language, it corresponds to an ideal case where frustration is as diluted as possible in space, while in disclinated structures it is concentrated along the defect lines (but in that case we get 'possible' structures). We use this value of $\bar{p}$ in order to get the approximate number of faces of the Voronoi polyhedron. The sites having three neighbours on each polyhedron, we get, from equation (4.3.9):

$$\sum_p (6 - p)F_p = 12 \tag{4.3.15}$$

Introducing $\bar{p} = (\sum pF_p)/F$ and assuming that the Voronoi cells are all equivalent one gets:

$$\bar{F} = 12/(6 - \bar{p}) \tag{4.3.16}$$

With the above value of $\bar{p}$, this gives $\bar{F} \simeq 13.39$.

It is now very interesting to look at the values that $\bar{F}$ takes in numerous examples. Figure 4.16 reproduces the compilation done by Nelson (1983). It is striking that the ideal 13.39 value falls in the same small range as the Frank–Kasper phases (see chapter 7). In fact one expects that any structure presenting icosahedral order in R^3 could be described as a flattened $\{3, 3, 5\}$ polytope with a disclination network. Such a structure will have an average coordination number very close to 13.39. It is tempting to relate deviation from the ideal value to geometrical and physical characteristics. Arguments along this line have been tentatively given in terms of the Voronoi cell asymmetry (Rivier 1982a) or of the homogeneity of the geometry of the disclination network (Sadoc and Mosseri 1984) (see also §5.4.3). As an example we describe in chapter 5 hierarchical polytopes in which the coordination tends to 13.33 in the flat limit.

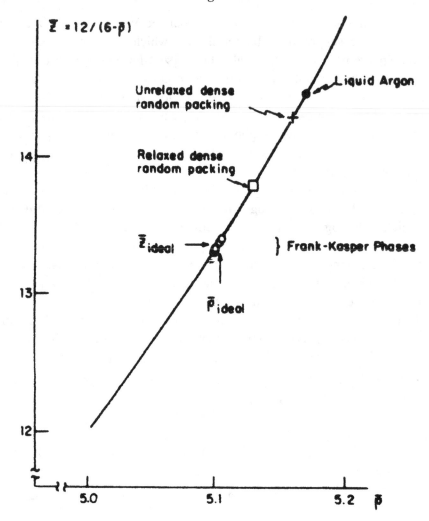

Fig. 4.16. Coordination number $\overline{Z}$ versus $\overline{p}$ for different kinds of dense atomic configurations.

4.3.4 Disclination density and Regge calculus

We are now trying to approximate the disclination density in three dimensions (Sadoc and Mosseri 1984). Recall first that in two dimensions, the problem is exactly solved. For a given coordination number, some polygons do not 'carry' curvature and are called 'neutral charges' (e.g., hexagons in three-fold coordinated two-dimensional tilings). As a result, on a unit radius spherical surface, summing the contributions of disclinations (identified to polygons different from the 'neutral charges') gives the area of the sphere. In this approach all the curvature is supposed to be concentrated at the centre of the polygons. There is a natural dual approach in which the curvature is concentrated on a vertex. This can be stated in terms of a deficit angle as in the approach proposed by Regge (1961), in order to discretize the curved manifolds of gravitation theory. The surface is assumed to be flat everywhere except on point defects or local concentrations of curvature. The surface is a polyhedron in an overall sense, since it is not necessarily a closed surface. In a simple case where faces are regular polygons, the angular deficit δ of each vertex is $\delta = 2\pi - q\alpha_p$ where α_p is the vertex angle of a p-gonal face. So,

$$\delta = 2\pi[4 - (p-2)]/(2p) \qquad (4.3.17)$$

The number of vertices in a regular polyhedron $\{p, q\}$ is deduced from the Euler relation

$$V = 4p/[4 - (p-2)(q-2)] \qquad (4.3.18)$$

Consequently the total angular deficit for a polyhedron is $\sum \delta = 4\pi$, the area of a unit radius sphere S^2.

Let us now consider the hypersphere S^3. Its three-dimensional volume (and not the hypervolume of its interior) is $2\pi^2 R^3$. Just as the usual sphere S^2 can be approximated by polyhedra, S^3 can be approximated by polytopes. Following Regge, the curvature is then concentrated on the edges (while it was concentrated at the vertices in two dimensions). To each edge it is possible to associate a cell, called an E-cell, which contains all points closer to this edge rather than to any other edge. This is a natural extension of the Voronoi cells associated with sites. In the case of a regular polytope $\{p, q, r\}$ the E-cells are regular r-gonal bipyramids. In less regular structures one can always construct the distorted bipyramid with, as base, the Voronoi cell face threaded by the considered edge. In the spherical space S^3, the base is a part of a geodesic spherical surface and its area can be calculated using the Gauss–Bonnet theorem. This area can be expressed in term of δ, the deficit angle along the edge:

$$\delta = 2\pi - r\theta \qquad (4.3.19)$$

θ is the dihedral angle of the cell $\{p, q\}$ of the polytope $\{p, q, r\}$. As a consequence, the volume V_E of the E-cell is given, within a good approximation if the edge length is short compared to the curvature radius, by:

$$V_E = l\delta r^2/3 \qquad (4.3.20)$$

where l is the geodesic edge length and r is the sphere radius. The total volume V of S^3 is then:

Table 4.1. *Evaluation of the S^3 volume, discretized by regular polytopes. s is the ratio of calculated volume V over the exact volume of the unit radius sphere. l is the geodesic length of the edges. As expected s tends to 1 when the number of edges increases*

$\{p, q, r\}$	δ	l	E	V	s
$\{2, 2, 2\}$	6.2831	3.1416	2	13.1594	0.667
$\{3, 3, 3\}$	2.5903	1.8234	10	15.7445	0.798
$\{4, 3, 3\}$	1.5707	1.0472	32	17.5459	0.889
$\{3, 3, 4\}$	1.3593	1.5707	24	17.0820	0.865
$\{3, 4, 3\}$	0.5513	1.0472	96	18.4737	0.936
$\{5, 3, 3\}$	0.1798	0.2709	1200	19.4902	0.987
$\{3, 3, 5\}$	0.1283	0.6283	720	19.3605	0.981

$$V = 2\pi^2 r^3 \simeq \sum_{edges} l\delta r^2 / 3 \qquad (4.3.21)$$

This relation is similar to:

$$\sum_{edges} l\delta = \frac{1}{2} \int \mathcal{R} g^{1/2} \, \mathrm{d}^3 x \qquad (4.3.22)$$

which appears in Regge calculus. $\mathcal{R}$ is the scalar intrinsic curvature (appendix A8) and g the metric tensor determinant. For an S^3 sphere $\mathcal{R} = 6/r^2$ and $\int g^{1/2} \, \mathrm{d}^3 x = 2\pi^2 r^3$.

Regge has developed his method in order to study discrete space-time, so he works using a Minkowskian metric, in place of a Riemannian one. But several results obtained by this method have a counterpart in our approach. For instance, conservation laws for disclinations which are presented in the next section (§4.4) in relation to Bianchi identities, are similar in a gravitational theory in $2 + 1$ dimensions, to the conservation laws of momentum (Roček and William 1985).

Regge calculus consists of a simplicial decomposition of the curved manifold and replacing integrals on the manifolds, by an integral on the simple skeleton. The space enclosed by a tetrahedron is regarded as flat and the curvature is concentrated on the edges. We can test the validity of this approximation by calculating the volume of S^3, discretized in the regular polytopes $\{p, q, r\}$. Note that here the polyhedral discretization is not necessarily tetrahedral. Table 4.1 gives the results, where s is the ratio of calculated volume, by (4.3.21), over the exact volume of the unit radius sphere S^3.

In the case of negative curvature, the E-cell volume can be approximated by:

$$V_E = |l\delta r^2|/3 \qquad (4.3.23)$$

as long as it is not too large. In this case the calculated volume is larger than the exact one.

So we have shown how a curved space can be faithfully represented by a Euclidean network (Regge skeleton). In terms of disclinations, in two-dimensional space, we had an exact relation:

$$\sum_i n_i \delta_i = \langle \kappa \rangle = \langle r^{-2} \rangle \qquad (4.3.24)$$

where $\langle \kappa \rangle$ is the average Gaussian curvature per unit area and n_i is the number of disclination points with angular deficit δ_i, per unit area. In three-dimensional curved space, we get an approximate relation

$$\sum_i L_i \delta_i \simeq 3 \langle r^{-2} \rangle \qquad (4.3.25)$$

since the scalar curvature $\mathscr{R}$ equals $6/r^2$. L_i is the disclination length (with angular deficit δ_i) per unit volume. It is now possible to evaluate the disclination density in a Euclidean close-packed structure. To such a structure there corresponds an underlying corrugated geometry with regions of positive and negative curvature (Gaspard *et al.* 1984). In the discretized Regge-like approach, curvature is concentrated on the edges, the curvature sign depending on the number of tetrahedra sharing the given edge. The requirement of vanishing curvature reads:

$$\sum_{i,i'} L_i^+ \delta_i^+ + L_{i'}^- \delta_{i'}^- \simeq 0 \qquad (4.3.26)$$

where L_i^+ and $L_{i'}^-$ are the lengths per unit volume of positive ($\delta_i^+ > 0$) and negative ($\delta_{i'}^- < 0$) disclinations. This would be exactly zero if disclinations were infinitesimal, e.g. if curvature was spread throughout the volume. In the present case the compensation of opposite curvature is approximate.

Let us proceed to a simple calculation in the case of the hierarchical polytope (in the case $\lambda = 3$) where we assume that all first neighbour distances are equal. These polytopes are described in chapter 5. If $\theta = 1.2306 rd$ is the tetrahedron dihedral angle, then, in the polytope P_i, each disclination edge carries a negative weight $\delta^- = 2\pi - 6\theta = -1.102 rd$, while the other polytope edges carry a positive weight $\delta^+ = 2\pi - 5\theta = 0.1283 rd$. Let us introduce the ratio $\Delta = L^+/L^-$. If the above relation was exact, then $\Delta = \delta^+/|\delta^-| = 8.589$, for the fully flattened polytope P_∞. In the iterative polytopes the Δ value is given by

$$\Delta^{(i)} = \frac{6(N_{12}^{(i)} + N_{14}^{(i)} + N_{16}^{(i)})}{N_{14}^{(i)} + 2N_{16}^{(i)}} \tag{4.3.27}$$

At the infinite limit, one finds $\Delta^{(\infty)} = 9$, which is rather close to the above ideal value.

4.4 Conservation laws

4.4.1 Bianchi identities

Let us first recall what the set of relations known as the Bianchi identities are; see for instance Misner *et al.* (1973), and de Witt and de Witt (1963). If R is the Riemann tensor, they read:

$$\frac{\partial R_{ikl}^n}{\partial x^m} + \frac{\partial R_{imk}^n}{\partial x^l} + \frac{\partial R_{ilm}^n}{\partial x^k} = 0 \tag{4.4.1}$$

These relations express in fact a conservation law for the curvature. Consider, in a three-dimensional Riemannian manifold, an infinitesimal cube with edges Δx_1, Δx_2, Δx_3 and the vectors normal to the cube faces pointing outside. A vector **A** defined at the origin of the cube is carried by parallel transport along a closed curve formed by the four edges of a square face (see figure 4.17). Consider the pair of faces defined by Δx_2, Δx_3 at $x_1 = 0$ and $x_1 = \Delta x_1$. On the first one the variation of A reads:

$$\delta A^\alpha = R_{\beta 23}^\alpha(x_1) A^\beta \Delta x_2 \Delta x_3$$

while it reads

$$-\delta A^\alpha = R_{\beta 23}^\alpha(x_1 + \Delta x_1) A^\beta \Delta x_2 \Delta x_3 \tag{4.4.2}$$

on the second one.

Note that if the **A** vector is transported successively around all the faces, one gets $\delta \mathbf{A} = 0$, because all edges are travelled an even number of times in opposite directions. The two opposite faces at $x_1 = 0$ and Δx_1 contribute for:

$$[R_{\beta 23}^\alpha(x_1 + \Delta x_1) - R_{\beta 23}^\alpha(x_1)] A^\beta \Delta x_2 \Delta x_3 = \frac{\partial R_{\beta 23}^\alpha}{\partial x_1} A^\beta \Delta x_1 \Delta x_2 \Delta x_3 \tag{4.4.3}$$

Combining equivalent equations for the two remaining pairs of faces leads to the above Bianchi identities.

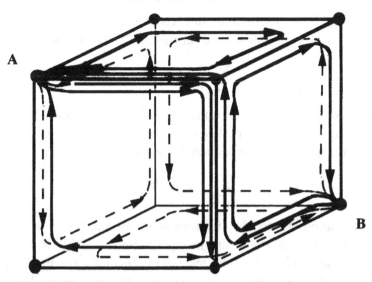

Fig. 4.17. Parallel transport along the edges of an infinitesimal cube. All edges are travelled an even number of times equally in both directions in order to come back to the starting point A.

4.4.2 Disclination equilibrium

Consider now a Riemannian manifold where all the curvature is concentrated on a network of disclination lines. Let us focus on a closed polyhedral surface surrounding a node of this network.

As in the analysis of Bianchi identities above, we study the parallel transport of a vector around the faces of the polyhedral surface (figure 4.18). Only faces

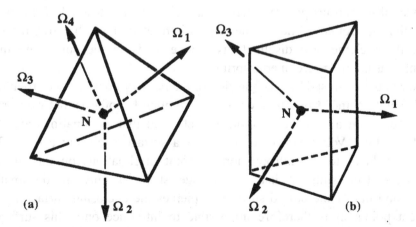

Fig. 4.18. A closed polyhedral surface surrounding a node in a disclination network: (a) a tetracoordinated node, (b) a tricoordinated node.

which are pierced by a disclination line contribute to rotating this vector. Assuming, in a first step, infinitesimal disclinations, the change in vector $\mathbf{A}$ is:

$$\delta \mathbf{A} = \sum_{disclinations\ i} \mathbf{\Omega}_i \times \mathbf{A} = 0 \qquad (4.4.4)$$

$\mathbf{\Omega}$ is a vector associated with the disclination: it is oriented along the disclination line, pointing outside of the closed polyhedral surface and its modulus equals the deficit angle of the disclination. It follows that

$$\sum_{disclinations\ i} \mathbf{\Omega}_i = 0 \qquad (4.4.5)$$

This is a conservation law for disclinations which means that, when isolated, they cannot bend and that nodes are in equilibrium as if disclinations were forces $\mathbf{\Omega}$. This is shown here for infinitesimal disclinations, but this properties extends to finite disclinations. As an example, consider three disclination segments of the same intensity, all belonging to the same plane. They are in equilibrium if they make a $2\pi/3$ angle. This is indeed the case for the Z_{15} sites in Frank–Kasper phases (see chapter 7 and appendix A6). The closed polyhedral surface surrounding the node can be taken as a triangular prism.

Consider a closed path defined by edges of this prism. Only circulation along the rectangular faces contributes to $\delta \mathbf{A}$, as disclinations pierce only these faces. In the case of infinitesimal disclinations the three rotations corresponding to the three rectangular faces are in equilibrium. Here, since we consider finite rotations, we are facing a difficulty due to the non-commutativity of rotations, with a paradoxical consequence suggesting that the space pierced by disclinations depends on the order among the disclinations. It is in fact possible to introduce disclinations progressively, as a series of infinitesimal disclinations in equilibrium around the node. The equilibrium relation which is true for each set of three elementary disclinations now extends to the sum. The three disclinations are therefore in equilibrium.

One can get confused by this question of the non-commutativity of rotations. The difficulty comes from the more or less implicit hypothesis that we start as a first step from a structure without defects, and that disclinations are then introduced by a Volterra process. Consider a surface enclosing a node. The problem reduces to the introduction of point disclinations on the surface. Consider first the simplest disclination node, such that only one disclination goes in and out of the closed surface. It pierces the enclosing surface at two different points: it is therefore impossible to introduce onto this surface a disclination at a single point. If several disclinations meets at a node, one gets just as many disclination points on the surface surrounding the node. The

Volterra process must be done simultaneously for all the point defects on the closed surface. This is the geometrical expression of the homogeneity needed in the introduction of disclinations. They have to be introduced progressively, as a sum of infinitesimal disclinations, and in a homogeneous way: an infinitesimal part of one after an infinitesimal part of another and so on.

The fact that a disclination cannot bend (make an angle on a node) can be shown more directly. Suppose that there is an angle, and take a vector **A** parallel to one of the two disclination segments. It is invariant if it circulates around one disclination and not around the other; but this is impossible since it must remain unchanged under a transport on a circuit drawn on a cylinder surrounding the two disclination segments.

Note that all these laws can be given in terms of a line tension, proportional to the angular deficit, such that all lines are in equilibrium.

4.5 Defects and topology

4.5.1 The homotopic classification of defects

A topologically stable defect is such that it cannot be suppressed (annealed) by a continuous distortion of the structure. A mathematical formalism, the homotopic classification of defects, has been developed in order to classify in a systematic way the topologically stable defects. It is not our aim to present it in details, but the interested reader can refer to original works or reviews (Toulouse and Kléman 1976, Toulouse 1980, Volovik and Mineev 1977, Mermin 1979, Trebin 1982, Venkataramann and Sahoo 1985, Venkataramann *et al.* 1989).

The main points of this theory are as follows. The system is described by a local order parameter, for instance a spin direction in a magnetic problem or the molecular orientation in the case of a nematic liquid crystal. All possible values taken by this parameter define an 'order parameter space', that we call V. Let us focus on linear defects in three dimensions (or point defects in two dimensions). To detect a defect amounts to surrounding its supposed location by a closed circuit. The evolution of the order parameter is followed in V along the image of this path. The topological characterization consists of counting all possible classes of topologically distinct closed paths on V, which corresponds to computing its first homotopy group, which therefore classifies point defects in two dimensions or linear defects in three dimensions. Other groups of higher order are also used: in three dimensions, second homotopy groups classify point defects, and third homotopy groups are related to 'textures'. This approach has been particularly fruitful in the study of continuous systems like liquid crystals. It can also be applied, but with some care (Mermin 1979), to atomic discrete structures: in the case of dislocations, the path defined above is related to the Burgers circuit, enclosed by its Burgers vector. An interesting property of the homotopic classification of defects is that the combination of defects can be traced back on the multiplication table of the homotopy group. If this group is non-Abelian, this implies topological obstructions to the crossing of defects. Nelson (1983) has used this property in his description of the dynamical properties of an 'icosahedratic phase' (see chapter 6).

4.5.2 2π and 4π disclinations in two dimensions

All the above properties vanish if the disclination angle is 2π: 2π disclinations can be bent, the rotation axis taking any direction relative to the disclination line. In two-dimensional space, it is possible to consider a 2π disclination as resulting from the sum of smaller disclinations: the core of the defect is spread. For instance four $\pi/2$ disclinations in a square tiling give globally a 2π disclination. Consider a two-dimensional space which has the topology of a cylinder, represented here as a cylinder of square (instead of circular) section, and tiled by squares with four squares sharing a vertex. Its Gaussian curvature vanishes, and so no disclination is carried. Now add at one end a 2π disclination, decomposed into four partial disclinations of $\pi/2$. This closes the end of the cylinder by a square, leading to four corners which are the partial disclinations (figure 4.19). More generally a cylinder closed by a half sphere is a flat space (cylinder) except in a finite region (half sphere) such that all the curvature is equivalent to a $+2\pi$ disclination. But this object would be more difficult to visualize if we supposed that the 2π disclination had collapsed into a single point.

Also, -2π disclinations are not too difficult to imagine in two dimensions, as shown by Harris (1977). For instance, two planes connected by a cut form a -2π disclination. Figure 4.20 shows an example of this kind of defect.

Let us now discuss another kind of defect, the $\pm 4\pi$ disclination, still more complex, but probably important in view of its relation to some fundamental (topological) properties of two- and three-dimensional spaces. A set of partial disclinations combining into a 4π disclination gives a two-dimensional space with the topology of a sphere (if the set collapses, only one point remains!). A -4π disclination can be generated by a standard Volterra construction, upon gathering three planes along cuts and in fact by introducing two -2π disclinations (figure 4.21a, b). But in a space that is flat almost everywhere, other possibilities arise for a -4π disclination. As shown in figure 4.21c, a tunnel connecting two planes is a good example. If we suppose that all the curvature

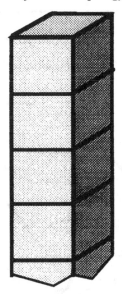

Fig. 4.19. A surface tiled by squares such that the set of disclinations is equivalent to a 2π disclination. This surface has the topology of a cylinder, infinite in one direction and closed at the other end.

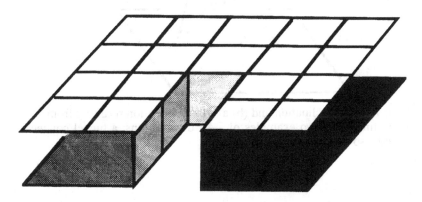

Fig. 4.20. A surface tiled by squares, such that its set of disclinations is equivalent to a -2π disclination.

is collapsed into a point, two planes with a common point are then a limiting case of this example (not that it is needed to embed the two planes at least in R^4 in order to avoid intersections).

A last example, showing that there are relations between genus and $\pm 4\pi$ disclinations, is a plane where a handle has been grafted (figure 4.21d). However, the exact relation between genus and $\pm 4\pi$ disclinations is not well understood, and should be studied further.

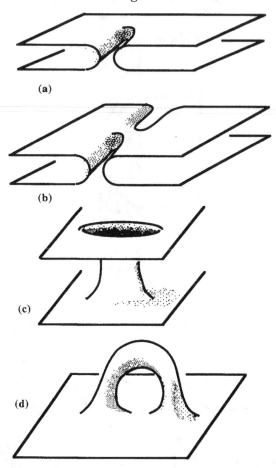

Fig. 4.21. (a) A -2π disclination and (b) a -4π disclination resulting from combining two -2π disclinations. Two examples of -4π disclinations: a tunnel connecting two planes (c) and a so-called 'worm hole' (d).

4.5.3 2π and 4π disclinations in three dimensions

Extension to three dimensions is done by adding a new dimension to the above point defects, which transform them into linear defects.

But there remain nevertheless many subtleties. The first one comes from the fact that a 2π disclination is not necessarily along a geodesic line. Looking at the above derivation of the Bianchi identities, shows that the conservation of the direction for a 2π disclination line is not required.

If we imagine a 2π disclination as a bundle of partial disclination lines, each line must respect its conservation rules, but globally the whole bundle can fluctuate in direction.

Consider an example in a Euclidean space tiled by cubes. In this space we take off a ring formed by several cubes (figure 4.22). So the space is now pierced by a hole having a border with a toroidal topology. We then build another space. A toroidal surface tiled by squares similar to the border of the above hole is extended to a third dimension in order to have a space with the topology $T^2 \times R$. Suppose that this space is tiled by cubes. If the torus happened to be a spherical torus, this tiling would have the same curvature as R^3. It is not the case here, because on the surface of the hole tiled by squares, there are sometimes three squares sharing a vertex and sometimes five, instead of always four, in a flat case. So this space with the topology of $T^2 \times R$ contains eight $\pi/2$ disclinations (with three cubes sharing an edge) and eight $-\pi/2$ disclinations (with five cubes sharing an edge) which are all parallel and balance each other. For the first time we can neglect these balanced disclinations.

Let us now cut this space into two parts limited by a toroidal surface, and then glue together one such half space and the Euclidean space, along its hole. We have then built a space containing four rings of $-\pi/2$ disclinations along the torus edges (shaded in figure 4.22). Indeed there are five cubes sharing such edges (three from the Euclidean pierced space, and two from $T^2 \times R$, which is locally equivalent to a Euclidean space). This set of four disclinations is equivalent to a -2π disclination which is closed on itself. The individual

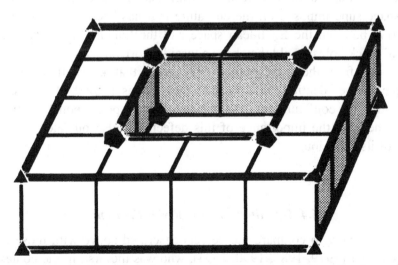

Fig. 4.22. A toroidal surface taken out of a Euclidean tiling of cubes contains a network of disclinations whose sum is a 2π disclination in a space obtained by adding a new dimension orthogonally to the surface. There are sometimes three and sometimes five squares around a vertex, displayed schematically by triangles and pentagons.

change in the direction of the partial $-\pi/2$ disclinations is due to the above neglected disclinations.

The parallel transport of a vector along a ring interlaced with the torus always indicates a -2π disclination whatever the precise orientation of the ring.

Even more complex are -4π disclinations in three dimensions. Indeed, it is still possible to add an extra dimension to all the examples given in two dimensions, but then more subtle configurations are possible.

For instance, consider the above example of the -2π disclination in a cubic lattice, but with two -2π disclinations. Then suppose that the two tubes obtained from $T^2 \times R$ are connected together. This gives a space looking like the one shown in figure 4.21d, but with one extra dimension, and with tori in place of circles in order to connect the 'handle'. This space has the same points at infinite distance as the Euclidean space, only a parallel transport along a loop interlaced with the two tori shows that it contains a -4π disclination.

This is a way to built some kind of 'worm hole' in three dimensions. Such an object has been introduced in general relativity with one more dimension, but it seems that the image which is given (in Euclidean space) in this context is closer to our two-dimensional example: two spherical holes connected by a 'handle'. This is really a -4π disclination in the two-dimensional example because the 'handle' has the same curvature as $S^1 \times R$, e.g. it has flat Euclidean metrics. But with one or two additional dimensions, the 'handle' is locally a $S^2 \times R$ space (or $S^3 \times R$ in four dimensions), which is already a very curved space superimposed on the curvature that has been introduced just at the junction between the Euclidean space and the 'handle'. So, what occurs with 'real' black holes: could they have a toroidal horizon?

It is known, from the study of the $SO(3)$ group, that 4π rotations in R^3 are homotopic to the identity, showing that they act as neutral elements. This should lead to the configuration of -4π disclinations spontaneously vanishing. We touch here on the importance of this defect, but also our lack of understanding of its behaviour.

4.5.4 Odd lines and the Rivier theorem

It is important now to compare the defects described above with those introduced by Rivier (1979, 1987, 1990, 1995), who was the first, in the context of glassy materials, to introduce disclinations, the latter being of a kind somehow different from those described above. Even if it seems still difficult to unify disclinations of our curved space model and 'Rivier lines' in the same approach, the parallel progress of the two points of view has certainly benefited from mutual conceptual exchanges.

The Rivier model begins with an elegant demonstration of a theorem on tricoordinated polyhedra, simply using the Euler relation. Rivier proves that the number of odd faces is even. This is a particular case of the relations given above. In a three-dimensional tetracoordinated structure, which is a packing of such polyhedra, it is then possible to follow continuous lines, which close onto themselves or end at the surface of the structure, and which thread odd rings only and avoid the even rings. Indeed each polyhedron has a pair of odd faces allowing a line to get in and out. Notice that the Rivier lines definition is ambiguous if the polyhedron has several pairs of odd faces, but each choice leads to a network of uninterrupted lines.

But tetracoordinated structures are not always made of cells, since they can contain twisted rings. Nevertheless, Rivier lines are still present, even if the demonstration is more difficult, using the concept of an irreducible ring. A ring is said to be irreducible if the shortest path between two vertices of a ring uses only edges of the ring. Then Rivier lines go through irreducible rings. Note that several have tried, unsuccessfully, to find counterexamples to the Rivier statement. A model builder, in the context of amorphous structures, believing them to work at random, found it difficult to accept that they followed in fact some invisible lines. But, as we have already seen in two dimensions, topology imposes its law to space tilers! In a recent work, Wooten (1995) proposes an algorithm allowing the automatic search for Rivier lines. We must emphasize an important difference between odd lines and disclinations: the first are topological objects while the second carry curvature and have to follow conservation rules related to the curvature of the space.

In parallel, Duffy and Rivier (1982) have proposed a gauge theory of glasses with the following hypotheses. The randomness of glasses leads to a lack of generic rotational or translational symmetry. Nevertheless glasses have a global isotropy and homogeneity such that every objective measure done on an atom could be similarly done on any other atom. It is this kind of symmetry we are facing, while being loose in a forest. Every atom, every tree, is as good as another to be the reference point, even if the local surroundings are different (tetrapod attached to each atom). The free energy of a glass is invariant if we transform the local surroundings of an atom into those of another (mainly by local rotation of the tetrapod), if that is done by adjusting the connection with the remaining system in a covariant way. The local rotation is then a gauge transformation. The glass homogeneity can be written in terms of a gauge invariance parametrized by the rotation group.

In a continuous medium elasticity approach, these authors have shown that among all classical defects (points, lines and walls), only linear defects are topologically stable. This is a direct consequence of the fact that the rotation group $SO(3)$ is not simply connected (see appendix A2, §2): $\pi_1(SO(3)) = Z_2$, the two-element group, so linear defects related to a 2π rotation (not shrinkable to the identity) are stabilized. These lines, taking account of the gauge invariance, are then associated to two-level systems (one such system for each line).

The correspondence with the above Rivier lines is then given, following a scheme close to the one we followed in the last section when we defined 2π and 4π disclinations. We suppose that the geometrical manifold where the disordered system 'lives' is a Euclidean space punctured by these lines (the cores of the disclinations are excluded from the manifold). Even rings are then so-called 'one-cycles', equivalent to the identity, odd rings becoming so only when circum-travelled twice. This last proposition results from the identification of line defects, the consequence of global homogeneity of the glass viewed as an elastic continuous medium, with geometric odd lines which, at a microscopic scale, go through the odd rings in a disordered network. Notice however that this identification is not a prerequisite for the validity of this gauge theory of glasses.

Is it possible to observe Rivier lines experimentally, even indirectly? The question is still open. Baskaran (1986) has suggested that they could influence electronic states in a way close to the effect of a magnetic field.

4.5.5 Wrong bonds in covalent networks

We noted in §2.5.4 that, in amorphous III-V alloys, even-membered rings are prevalent as a result of chemical order, in order to allow for heteropolar bonds in the material. But a finite density of homopolar bonds can occur, called 'wrong bonds', resulting for instance from a local stoichiometry variation (Gheorgiu and Theye 1981). It is then interesting to study the way these wrong bonds are arranged in the structure (Mosseri and Gaspard 1987). Consider as a first step the case of an even structure, and put a wrong bond on a ring. It seems obvious that this ring must contain another wrong bond. Indeed, let us identify the two sites of the homopolar bond, and try to propagate a heteropolar order in the ring. The identification has changed the even ring into an odd ring, which cannot accommodate alternation of two kinds of site. Therefore, each even ring must contain an even number of wrong bonds. Using a reasoning similar to that used for Rivier lines, we conclude that wrong bonds are on lines which can never end in the structure, either closing or ending at the surface. This is illustrated in figure 4.23 in two dimensions.

But these wrong bonds cost energy, which would require to have them as

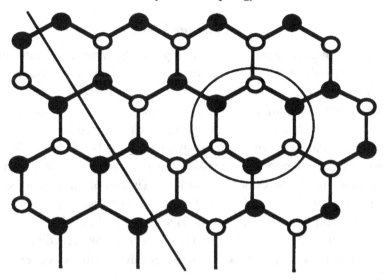

Fig. 4.23. Wrong bonds in two dimensions: they are organized along lines which either end at the surface or close.

isolated as possible. This can be done in an interesting way in disordered structures by allowing for odd rings. The latter can act as sources, or sinks for the wrong bond lines. In three dimensions adjacent odd rings can 'pin' a wrong bond, as shown in figure 4.24. Note that this configuration of odd rings is threaded by a Rivier closed line, showing an interesting duality between wrong bonds and Rivier lines.

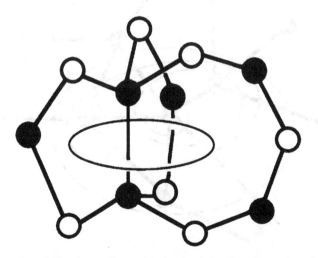

Fig. 4.24. Wrong bonds in three dimensions: pinning of a wrong bond by odd rings. The wrong bond is surrounded by a Rivier line.

4.5.6 Dangling bonds

In amorphous covalent networks, dangling bonds have important effects, mainly on electronic properties. In amorphous silicon for instance, these dangling bonds introduce new states in the gap, and alter the semiconducting properties. The saturation of these dangling bonds with hydrogen atoms opened a large field for amorphous semiconductor research at the end of the 1970s.

Is it possible to characterize the topology of dangling bonds as is done above for wrong bonds? It is a rather more difficult problem, and we only present here some indications in this direction. The structure (in two or three dimensions) is taken as a graph, with sites of variable coordinance and any type of rings. Consider a site, and build (see figure 4.25) successive shells B_l surrounding it in such a way that each new shell adds one new site to the preceding shells. So such a shell contains l sites. Let us enumerate A_l, the number of edges crossing a shell. We get

$$A_l = \left(\sum_{i=1}^{l} c_i \right) - 2(l + R_l - 1) \tag{4.5.1}$$

where c_i is the site coordinance and R_l the cyclomatic number, also called the first Betti number of the graph, which counts the number of irreducible rings inside the shell B_l. In the example of figure 4.25, we have

$$A_1 = 5, \ A_2 = 7, \ A_3 = 9, \ A_4 = 8$$

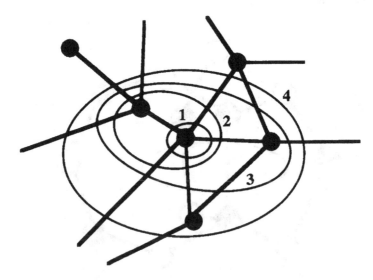

Fig. 4.25. A scheme defining shells around sites of given coordinance. Each new shell adds one more site to the preceding ones.

For a plane tiling, R is simply related to the number of faces F inside the shell, being F for a plane and $F - 1$ on a sphere where the last shell collapses to a point. In three dimensions, this number is more difficult to calculate, but it plays a crucial role characterizing the associated graph. We consider a structure with N sites in an n-dimensional space without border, which is divided into C cells intercepting edges and never passing through a site. By summation the above relation gives

$$\sum_{l=1}^{C} A_l = \sum_{j=1}^{N} c_i - \sum_{l=1}^{C} 2(l + R_l - 1)$$

This is twice the number of edges cutting all shells. We can then conclude that the sum of all coordinances is even:

$$\sum_{j=1}^{N} c_j = 2M \tag{4.5.2}$$

where M is an integer. This implies that the number of odd-coordinated sites is even.

Consider now in S^3, a tetravalent structure, containing therefore even-coordinated sites, except for a few 'odd' sites corresponding to dangling bonds. So, for a topological reason, it cannot contain an isolated single dangling bond, but at least two. If we consider then a structure in R^3 having a dangling bond somewhere near its centre, its presence will influence the parity of the number of dangling bonds at the surface (even very far away from the dangling bond itself). In fact, more generally, the parity of dangling bonds at the surface gives us information about the parity of dangling bonds in the bulk.

5

Hierarchical polytopes

5.1 Hierarchical polytopes and symmetry groups

5.1.1 Symmetry group of the {3, 3, 5} polytope and homotopy theory of defects

Line defects in condensed media can be classified using the homotopy theory (Toulouse and Kléman 1976), which allows us to identify the possible defects and to analyse their mutual relationships (Rivier and Duffy 1982). In the polytope {3, 3, 5}, linear defects are given by the conjugacy classes of the fundamental group $\pi_1(SO(4)/G')$ where G' is the [3, 3, 5] direct symmetry group of the polytope (Nelson and Widom 1984).

The polytope full symmetry group is presented in appendix A2, §3 using quaternions and in appendix 4 using the orthoscheme description. It contains 14 400 elements while the direct group G' contains 7200 elements. It is related to the icosahedral group Y by: $G' = Y' \times Y'/Z_2$, where Z_2 is a two element group and Y', of order 120, is the lift of the R^3 icosahedral group Y into $SU(2)$ ($SU(2)$ is isomorphic to the group of unit quaternions). This shows that elements of G' can be described as a combination of screws (sometimes called 'double rotations'). Related defects also combine several screws along different axes, leading to 'dispirations' similar to those introduced by Harris (see §4.1).

Among these defects are those related to $2\pi/5$ rotations. The introduction of two interlaced disclinations of this kind into the polytope {3, 3, 5} has already been presented in chapter 4. It changes the 120 vertices of the polytope into 144 vertices of type Z_{12} and 24 of type Z_{14}. If space curvature is measured in terms of the mean distance between vertices, then the global intrinsic curvature has been slightly decreased. One would like to iterate this procedure and eventually to achieve the complete flattening of the structure. The final model would then consist of regions of positive curvature, where the local geometry of the polytope is maintained, and of negative curvature concentrated on line defects, arranged in such a way as to balance the positive curvature.

We are then facing a major difficulty: a step by step incorporation of defects proves to be highly complex, due to the non-commutative character of the rotations. A method of introducing a finite number of disclinations (with the help of the Hopf fibration) was presented in chapter 4, but this approach could not reach a full decurving. It is however possible to by-pass this difficulty and achieve a complete flattening of the polytope and get an infinite structure in R^3. The key idea, which is now described, consists of introducing at each step not of a single disclination line, but a complete disclination network, whose symmetry group is contained in G (Mosseri and Sadoc 1984, Sadoc and Mosseri 1985a).

5.1.2 Hierarchical disclination networks

The most precise way to describe the hierarchical disclination networks is to use a deflation of the orthoscheme tetrahedron in the polytope. This method is described in §5.5.

But it is also possible to describe, more simply, the same disclination configuration by a decoration procedure of the tetrahedral cells of the successive polytopes. This decoration is chosen here in such a way as to generate dense structures at each step, while keeping a large amount of local icosahedral order. After each decoration step, a rescaling is done in order to keep a constant average first neighbour distance. Consequently, the radius of the hypersphere containing the polytopes increases upon iteration, so the space curvature is decreased.

5.1.3 Friauf–Laves decoration

Consider a tetrahedral cell in the $\{3, 3, 5\}$ polytope and add two new vertices on each of its edges, dividing them into three equal segments. The solid tetrahedron has been decomposed into four smaller tetrahedra and one truncated tetrahedron. This truncated tetrahedron is an important structural unit observed in several metallic compounds (Samson 1968). It is often called the Friauf–Laves polyhedron by crystallographers (figure 5.1).

All $\{3, 3, 5\}$ cells are thus decorated, leading to a decomposition of the polytope into tetrahedral and Friauf–Laves cells. New vertices are then added at the centre of Friauf–Laves polyhedra. There are three types of sites in this curved structure:

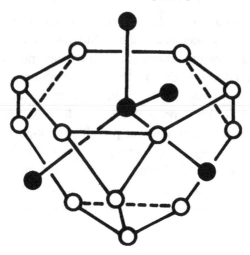

Fig. 5.1. A Friauf–Laves polyhedron. It is a truncated tetrahedron: its faces are equilateral triangles or hexagons. Its vertices, denoted M_1 in the text, are represented by open circles; black circles represent Z_{16} sites placed at the centre of the Friauf–Laves polyhedra.

- those located at the $\{3, 3, 5\}$ vertices, called M_0;
- those added (by pairs) on each $\{3, 3, 5\}$ edge, called M_1;
- and finally those added at the centre of the $\{3, 3, 5\}$ cell (therefore at the centre of the Friauf–Laves polyhedra), called M_3.

The proposed hierarchical flattening procedure is based on the iterative transformation of a tetrahedrized structure into another tetrahedrized structure containing more tetrahedra and vertices. From the set of new vertices M_0, M_1 and M_3, we define new tetrahedral cells, which are smaller than the initial cells. Let us enumerate these new tetrahedral cells. Each hexagonal face of a Friauf–Laves polyhedron carries six tetrahedra (shared by two neighbouring Friauf–Laves polyhedra). In addition, one finds four tetrahedra connecting the centre of a Friauf–Laves polyhedron to the four Friauf–Laves polyhedra triangular faces. Finally, one is left with four new tetrahedra capping these four triangles. Therefore, upon weighting the shared cells by one half, we see that a $\{3, 3, 5\}$ tetrahedral cell is decomposed into 20 smaller cells.

The coordination polyhedron of an M_0 site is an icosahedron formed by twelve M_1 sites (M_0 is therefore a Z_{12} site). An M_1 site is surrounded by one M_0, one M_1 site on the same edge (denoted M_1' in figure 5.2), five M_3 sites (at the centres of the five cells sharing the E edge), and finally five M_1 sites on the five edges adjacent to E, and coincident with E on M_0' (figure 5.2). It

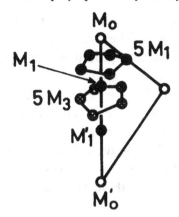

Fig. 5.2. Relation between points in the Friauf–Laves decoration of the $\{3, 3, 5\}$ polytope. M_0 are vertices of the $\{3, 3, 5\}$. M_1 are vertices of the Friauf–Laves polyhedra (or points on the edges of $\{3, 3, 5\}$). An M_0 vertex is surrounded by twelve M_1 vertices; an M_1 vertex is surrounded by one M_0, five M_3 and six M_1; an M_3 vertex is surrounded by twelve M_1 and four M_3.

follows that an M_1 site is also surrounded by an icosahedral coordination polyhedron.

M_3 sites have sixteen neighbours (they are Z_{16} sites), which are the twelve M_1 sites on the tetrahedral edges (Friauf–Laves polyhedron vertices) and the four M_3 sites at the centre of the four neighbouring Friauf–Laves polyhedra. Disclination lines go through the hexagonal faces of the Friauf–Laves polyhedra: they connect the centres of the $\{3, 3, 5\}$ cells, the M_3 sites, so that the disclination network is formed by the edges of the dual $\{5, 3, 3\}$ polytope. Recall that polytope $\{3, 3, 5\}$ has 120 vertices, all of which are twelve-fold coordinated. The above decoration yields a new structure containing 1560 atoms twelve-fold coordinated (Z_{12}) and 600 atoms on Z_{16} sites. This struc-ture is called the P_1 polytope and its main content is summarized in table 5.1. The different coordination polyhedra have only triangular faces; consequently the P_1 polytope is a packing of tetrahedral cells, five or six sharing a given edge.

The decoration has been described for one regular tetrahedral cell of the $\{3, 3, 5\}$ polytope, but nothing prevents us from applying it again to a P_1 cell even if these tetrahedra are not regular. This is the starting point of the iterative method generating hierarchical defects. The resulting structure (a P_2 polytope) is very similar to P_1, except that a new kind of coordination polyhedron now appears: Z_{14} sites. These sites correspond to vertices added on to an edge common to six tetrahedral cells of the P_1 polytope, which are now decorated. Thus, Z_{14} sites appear in-between two previous Z_{16} sites on disclination segments of the P_1 polytope. Polytope P_2 contains two dislocation networks, disjoint and interlaced in a way schematized in figure 5.4. It is possible to iterate *ad infinitum* this decoration procedure and obtain larger and larger structures. At each iteration, a scaling factor $\lambda = 3$ is generated between each P_i polytope and, consequently, between each interlaced disclination network. Pairs of Z_{14} sites are added between previous disclinated sites (Z_{14} and Z_{16}).

There is another way to describe the same decoration using the Voronoi cell surrounding a given vertex. The Voronoi cell for a vertex of the $\{3, 3, 5\}$ is a dodecahedron $\{5, 3\}$, the basic cell of the dual polytope $\{5, 3, 3\}$. A new polytope is generated by filling in these Voronoi cells with suitably oriented centred icosahedra (figure 5.3).

We noted above that P_1 has 2160 vertices, distributed as follows: there are 600 vertices on the Voronoi cell vertices (the number of sites in the $\{5, 3, 3\}$). Each of the 120 dodecahedral cells is filled with thirteen vertices (the centred icosahedron) which add up to the remaining 1560 sites. The transformation can be repeated by considering the Voronoi decomposition of the structure and filling these cells again with appropriate centred deltahedra. At the second step, Z_{14} sites are generated. When the process is repeated again only Z_{12}, Z_{14} and Z_{16} are generated.

After p iterations, polytope P_p contains p interlaced disclination networks, which all share the [3, 3, 5] symmetry group. As a consequence, these hierarchical structures display orientational order. This has been tested both numerically and by optical Fourier transform (Mosseri and Sadoc 1982, Sadoc 1985)

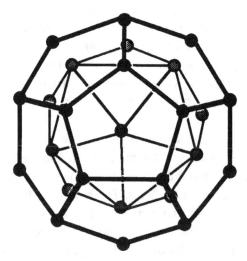

Fig. 5.3. Decoration of the dodecahedral Voronoi cell by a centred icosahedron.

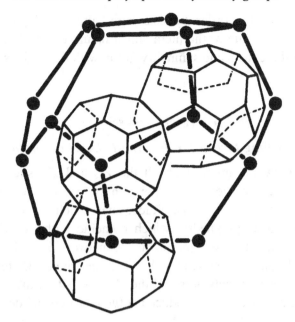

Fig. 5.4. Local view of two interlaced disclination networks corresponding to two successive decurving operations. A dodecahedron, which is the elementary cell of one network ({5, 3, 3} edges), is represented by heavy lines. A second network is shown interlaced with the first one.

of large clusters, mapped onto a tangent R^3 after several iterations. Figure 5.5 shows the optical Fourier transform along five-fold and three-fold axes for one such cluster. Note that in the disclination mediated theory of two-dimensional melting, disclinations destroy the orientational order (Nelson and Halperin

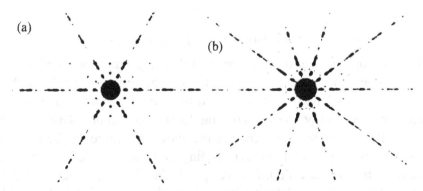

Fig. 5.5. Two-dimensional optical Fourier transform of a finite piece of the hierarchical polytope P_2 orthogonally mapped onto a tangent R^3: (a) three-fold symmetry axis; (b) five-fold symmetry axis.

1979). The fact that this order persists here, although highly disclinated networks are present, is due to the hierarchical nature of the defect networks, which in addition share the same symmetry group with the initial structure.

5.2 Hierarchy and scaling

5.2.1 Scaling factors for the Friauf–Laves decoration

The disclination networks go through all the Z_{14} and Z_{16} sites. At the first iteration the Z_{16} sites are connected along the edges of the dual polytope $\{5, 3, 3\}$, since its nodes (Z_{16} sites) are those sites which are introduced at the centres of Friauf–Laves polyhedra, which coincide with the $\{3, 3, 5\}$ cell centres. At the second iteration, the preceding network is still present (with greater edge length) and is interlaced with a new one having shorter edges. The Z_{16} sites are nodes in both networks and the Z_{14} sites sit on the edges (larger) of the first disclination network, introduced at the first step and decorated at the second.

The new defect network introduced at each iteration is formed by the edges of the dual of the previous polytope, and the topology of the disclination networks obtained at all previous iterations remains unaffected. Upon iteration, interlaced tetravalent disclination networks are generated. Each one has its own length scale which is the distance between two nodes in the network (Z_{16} sites). In the structure obtained at iteration i the network introduced by a previous iteration j has an internode distance equal to 3^{i-j}, in intersite distance units. Note that this scaling factor is only approximate since most of the new tetrahedra are not regular.

5.2.2 Other scaling factors

The above decoration has been described by stating that, upon dividing the edges into three equal parts ($\lambda = 3$), each $\{3, 3, 5\}$ tetrahedral cell has been cut into one Friauf–Laves polyhedron and four smaller tetrahedra. But there are many other distinct types of interesting decorations, among which we have described only the simplest one. Indeed, the above procedure can be extended to any scaling with λ an odd number, leading to the splitting of a tetrahedron into smaller tetrahedra and Friauf–Laves polyhedra. To demonstrate this statement, consider a large equilateral triangle (edge length c) tiled with triangles and hexagons (edge length $a = c/\lambda$, with $\lambda = 2k + 1$). This pattern is sometimes called a Kagome tiling. An example is shown in figure 5.6, with $\lambda = 5$.

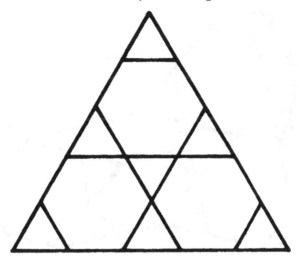

Fig. 5.6. Decoration of a tetrahedron face in the case $\lambda = 5$.

This large triangle is a face of the tetrahedron which we aim to divide into tetrahedra and Friauf–Laves polyhedra.

In order to describe the decoration of a tetrahedral cell, we pile up Friauf–Laves polyhedra and tetrahedra on top of the large triangle in order to form a large tetrahedron; we refer the reader to some of the figures of the book written by William (1979). On each hexagon, put a Friauf–Laves polyhedron such that its triangular face opposite the hexagon has the same orientation as the large triangle. (There are three such Friauf–Laves polyhedra when $\lambda = 5$.) There is room for other Friauf–Laves polyhedra, sharing now a triangular face with the tiling (only one triangle, located at the centre, when $\lambda = 5$). These new Friauf–Laves polyhedra share some hexagonal faces with the Friauf–Laves polyhedra already in place. Small tetrahedra finally fill in the remaining holes. A layer of the Friauf–Laves polyhedra and tetrahedra is therefore obtained. The upper face of this layer is an equilateral triangle of reduced edge length $(\lambda - 2)a$ (this is $3a$ with $\lambda = 5$), again tiled by a Kagome pattern. It is then possible to iterate the covering of these upper triangles by new layers of Friauf–Laves polyhedra and tetrahedra, up to filling a large tetrahedron of side c. Figure 5.7 shows some Friauf–Laves polyhedra organized according to this rule. In figure 5.8, decoration of tetrahedra for different λ values is presented.

By putting atoms on the vertices and at the centres of every Friauf–Laves polyhedron, a new close-packed structure is obtained in curved space. Note that the well-known crystalline cubic Laves phase corresponds to a stack of infinite layers formed by Friauf–Laves polyhedra and tetrahedra. The decorated tetrahedron can therefore be considered as part of a cubic Laves crystal. Thus,

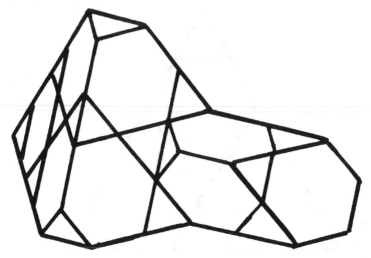

Fig. 5.7. Regular packing of Friauf–Laves polyhedra and tetrahedra.

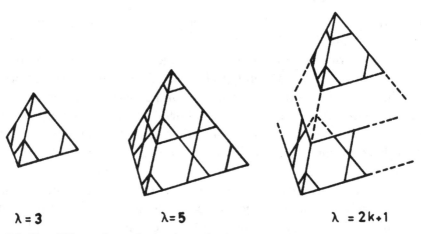

$$\lambda = 3 \qquad\qquad \lambda = 5 \qquad\qquad \lambda = 2k+1$$

Fig. 5.8. The filling of tetrahedra by Friauf–Laves polyhedra and tetrahedra. The scaling factors are $\lambda = 3, 5$ and $2k + 1$.

we have obtained an infinitely large family of possible decorations parametrized by the odd integer λ, the case $\lambda = 1$ obviously corresponding to the identity transformation.

5.2.3 Disclination lines in the structure obtained by decorating the {3, 3, 5} cells by Friauf–Laves polyhedra

We have seen above that, for $\lambda = 3$, the first decorated structure is a polytope (P_1), whose disclination network is formed by {5, 3, 3} edges. When λ is greater than 3, one also obtains a disclination network, but with a different geometry (figure 5.9).

Sites inside a tetrahedral cell have exactly the same local environment as atoms in Laves phases. Sites on the Friauf–Laves polyhedra vertices have icosahedral coordination shells (Z_{12}); sites at the centre of Friauf–Laves polyhedra are 16-fold coordinated (Z_{16}), and sites on the vertices, on the faces or on the edges of the tetrahedral cells are Z_{12} sites (the latter being also vertices of Friauf–Laves polyhedra). Consequently, the disclination network is formed by segments connecting the centres of the Friauf–Laves polyhedra (Z_{16}) through their hexagonal faces; it is therefore tetracoordinated. In order to understand better this defect network geometry, one can split it into the part which is contained inside the large tetrahedral cells and the part connecting these cells. This disclination network inside the $\{3, 3, 5\}$ cells has the same bond topology as in a diamond structure (like the Frank–Kasper lines in the cubic Laves phases; see chapter 7). Around the vertices of the $\{3, 3, 5\}$, the disclination network forms dodecahedral cages; around its edges, it makes so-called large barrelans (figure 5.10). Thus, every tetrahedral cell of $\{3, 3, 5\}$ is

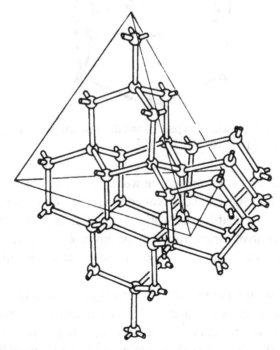

Fig. 5.9. Disclination lines in the structure obtained by decorating the $\{3, 3, 5\}$ tetrahedral cells by Friauf–Laves polyhedra. Only Z_{16} sites and the disclination lines connecting them are shown.

Table 5.1. *Information about the iterative polytopes P_i ($\lambda = 3$): N_p is the
number of p-fold coordinated vertices, N the total number of vertices, T the
number of tetrahedra and $\bar{z}$ the mean coordination number*

	N_{12}	N_{14}	N_{16}	N	T	$\bar{z}$
P_0	120	0	0	120	600	12
P_1	1560	0	600	2160	12 000	13.111
P_2	27 480	2400	12 600	42 480	240 000	13.299
P_3	537 240	57 600	252 600	847 440	4 800 000	13.328
P_4	10 706 520	1 183 200	5 052 600	16 942 320	96 000 000	13.332
P_5	214 014 360	23 760 000	101 052 600	338 826 960	1 920 000 000	13.333

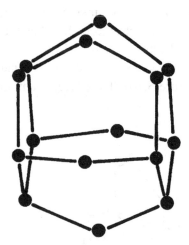

Fig. 5.10. A large 'barrelan', which is part of the disclination network, appears around
the edges of the $\{3, 3, 5\}$ polytope when $\lambda = 3$.

naturally filled by a four-coordinated network, which is diamond-like inside,
barrelan shaped on the edges and dodecahedral around the vertices.

Note that these disclination networks have similar topologies to those
used to describe covalent structures. For example, $\lambda = 3$ corresponds to the
$\{5, 3, 3\}$, and other values of λ correspond to local order similar to those
presented in §3.1.7.

In principle, an infinite number of iterations is required to completely flatten
the polytope. However, as seen in table 5.1, the number of vertices increases
greatly upon iteration. In order to compare the successive structures, one
should rescale the S^3 radius in order to get a constant first neighbour distance.
This implies that, after a limited number of iterations, this radius is so large
that huge, almost flat, models can be obtained on R^3 by orthogonal mapping of

the small neighbourhood surrounding the point of contact between S^3 and the tangent space R^3.

5.3 Matrix formulation of the hierarchical structures

5.3.1 Matrix formulation: the $\lambda = 3$ example

We have seen how the hierarchical structure can be described geometrically by the disclination network; in this section we use an algebraic approach which leads to the numbers of all Z_{12}, Z_{14} and Z_{16} sites at each step of the decoration (Sadoc and Mosseri 1985a). Let us gather the three numbers n_{12}, n_{14} and n_{16} of Z_{12}, Z_{14} and Z_{16} sites after the ith iteration into a three-dimensional vector $\mathcal{N}^{(i)}$. We can then write a matrix Ω, which acts as a transfer matrix, and relates the representative vectors of the structure before and after one decurving operation:

$$\mathcal{N}^{(i)} = \Omega \mathcal{N}^{(i-1)} \tag{5.3.1}$$

In order to build the transfer matrix corresponding to the decurving operation $\lambda = 3$, consider the decomposition of the Voronoi cells and their filling with appropriate centred deltahedra Z_{12}, Z_{14} and Z_{16}. When the process is repeated indefinitely, only these three types of coordination shell are generated. The transformation can be written formally:

$$Z_{12} \rightarrow 13Z_{12} \qquad\qquad + 5Z_{16}$$

$$Z_{14} \rightarrow 12Z_{12} + 3Z_{14} + 6Z_{16} \tag{5.3.2}$$

$$Z_{16} \rightarrow 12Z_{12} + 4Z_{14} + 8Z_{16}$$

The Ω matrix thus reads

$$\Omega = \begin{pmatrix} 13 & 12 & 12 \\ 0 & 3 & 4 \\ 5 & 6 & 8 \end{pmatrix} \tag{5.3.3}$$

The iteration begins with the $\{3, 3, 5\}$ and so $\mathcal{N}^{(0)} = (120, 0, 0)$. Multiplication by Ω gives information about the successive polytopes as summarized in table 5.1.

To the largest eigenvalue of Ω (the so-called Perron root) corresponds an eigenvector whose components give the average properties of P_∞. Indeed, it refers to an asymptotic situation where the relative fraction of different types of site remains constant under iteration. In the present case the Perron root is $v = 20$ and the P_∞ mean coordination number, $40/3$, is closely approached after only very few iterations. This number is also not far from that of the Coxeter 'statistical honeycomb' discussed in other chapters (§4.3.3, A4.2.2).

5.3.2 Matrix formulation for any odd value of λ

When the above procedure is extended with a decoration associated with a generic odd value for the scaling parameter λ, the transfer matrix reads

$$\Omega_\lambda = \begin{pmatrix} 10A + \lambda & 12A & 14A - (\lambda - 1) \\ 0 & \lambda & 2(\lambda - 1) \\ 5A & 6A & 7A + 1 \end{pmatrix} \qquad (5.3.4)$$

with $A = \lambda(\lambda^2 - 1)/24$.

It is easy to check that $\Omega_1 = I$, the identity, as it should, and that Ω_3 has the form obtained in the previous section. Checking for higher values of λ becomes fairly laborious. It is easier to notice first that $A(\lambda)$ satisfies the recursion relation

$$A(\lambda) = A(\lambda - 2) + \frac{1}{4}(\lambda - 1)^2 \qquad (5.3.5)$$

The matrix Ω_λ has the general form

$$\Omega_\lambda = \lambda\mathscr{I} + (1 - \lambda)\mathscr{X} + 17A(\lambda)\mathscr{Z}$$

The basic matrices $\mathscr{I}$, $\mathscr{X}$, $\mathscr{Z}$ and $[\mathscr{X}, \mathscr{Z}]$, which are defined in equation (5.4.2), form a closed set under multiplication, as shown below. The three Ω_λ eigenvalues are 1, λ and $v(\lambda) = 17A + \lambda$, the latter being the Perron root; the eigenvalues correspond therefore to scales of $\lambda^0 = 1$, λ^1 (length) and λ^3 (volume) respectively. The Perron eigenvalue gives the multiplicative factor for the number of tetrahedra upon iteration (and therefore decurving) by Ω. Note that $v(1) = 1$, $v(3) = 20$, $v(5) = 90$, $\ldots$, so that $\lambda/v \ll 1$ and the Perron root dominates the decurving at every stage. The eigenvectors are

$$\mathbf{I} = \begin{pmatrix} 1 \\ -2 \\ 1 \end{pmatrix}, \quad \mathbf{\Lambda} = \begin{pmatrix} 6 \\ -5 \\ 0 \end{pmatrix}, \quad \mathbf{V} = \begin{pmatrix} 2 + \chi \\ 2\chi \\ 1 \end{pmatrix}$$

where $\chi = 24/[17\lambda(\lambda + 1)]$. Note that these vectors are not orthogonal, because the matrix is not Hermitian. From $\mathbf{V}$ it is possible to obtain a mean coordination number $\bar{z} = 40/3$, independent of λ.

5.3.3 Generating other Z_n

Here we have focused on an example of decurving operations generating only Z_{12}, Z_{14} and Z_{16} for any value of the scaling factor λ. It is also possible to generate hierarchical polytopes with other Z_n sites. One example is described elsewhere (Sadoc and Mosseri 1985a) which generates Z_{18} sites. The corresponding transfer matrix is:

$$\Omega = \begin{pmatrix} 45 & 48 & 56 \\ 0 & 5 & 12 \\ 10 & 12 & 17 \end{pmatrix} \tag{5.3.6}$$

acting on a vector: $(n_{12}^i, n_{14}^i, n_{18}^i)$. The Perron eigenvalue is 61, and the mean coordination number $\bar{z} = 13.2$.

5.4 Disorder and non-commutative defects

Up to now, we have considered only a simple type of iteration which is repeated endlessly, leading to a hierarchical structure with an asymptotic self-similar property. It is very interesting to devise more random processes, in the context of amorphous structure description, for instance. The procedure will still be repeated until decurving is complete, but a random choice among different transformations (different λ values) will be allowed at each step. Note that another possibility, generating even more disorder, would be to randomly alternate decorations which give rise to different Z_n types of sites. This will not be considered in the following. We shall here focus on mixing Friauf–Laves like decorations by varying λ.

It is possible to give a quantitative analysis of the derived structures and even to describe the most probable configuration in a statistical approach (Sadoc and Rivier 1987). The amount of disorder can thus be measured (or monitored, in a model builder's point of view) at a higher level, the level describing the nature of the disclination network as opposed to the atomic scale level, according to the nature of the sequences of iterations, either periodic, quasiperiodic or random.

But, as mentioned above, such structures are still quite ordered. To introduce more disorder, we shall consider a decurving operation that depends on the local position and leads to a non-uniform spatial disordering. This will generate more defects than those needed to decurve uniformly, mainly in the form of extra positive disclinations, and the resulting structures are believed to give more realistic glassy structures.

5.4.1 Disorder in the sequence of decurving processes

The total number N of atoms, and the total length of disclinations (which is in a first order approximation proportional to the total elastic energy E) are proportional to trace($\prod_i \Omega_i$) or, as we have seen above, to the scaling in volume, since Perron roots $v_i(\lambda)$ dominate the asymptotic behaviour. At every stage, decurving can be done in several ways, corresponding to the introduction of more or less violent disclinations. These options are parametrized by the odd integer λ, which measures the division of the length scale in the hierarchical procedure, $\lambda = 3$ corresponding to the most gentle decurving.

5.4.2 Algebraic description of the family of aperiodic flat structures

There are many possible infinite sequences of decurving matrices characterized by the sequence of λ. Random sequences $\{\lambda_i\}$ generate an infinite number of structures that are flat, hierarchical (by construction), aperiodic, but not locally disordered, because the structures are given globally by a single algorithm, the sequence $\{\lambda_i\}$ (which may be infinite). The configuration entropy of a structure goes only as w, the number of operations (approximately $N^{1/3}$) and not as N as it would be for a strong disorder. One can show by induction that all these matrices, including $\prod_i \Omega_i$, have the form

$$\prod_i^w \Omega_i = \left(\prod_i \lambda_i\right)\mathscr{T} + \left(1 - \prod_i \lambda_i\right)\mathscr{K} + \left(\prod_i v_i - \prod_i \lambda_i\right)\mathscr{Z}$$

$$+ \left(\prod_i \lambda_i - 1 - \Delta\right)\mathscr{C} \tag{5.4.1}$$

where $\prod_i \lambda_i$ is the product of λ_i, $\prod_i v_i$ is the product of $v(\lambda_i)$ and

$$\Delta = \cdots (\lambda_l - 1)\prod_k^{l-1} v_k + (\lambda_{l-1} - 1)\prod_k^{l-2} v_k + \cdots + (\lambda_2 - 1)v_1 + (\lambda_1 - 1)$$

$$\tag{5.4.2}$$

is the only coefficient depending on the ordering of the decurving, i.e. of the order of the sequence $\{\lambda_i\}$. For a single decurving, $\Delta = 0$. The four basic matrices:

$$\mathscr{T} = \begin{pmatrix} 1 & 0 & 0 \\ 0 & 1 & 0 \\ 0 & 0 & 1 \end{pmatrix}$$

$$\mathscr{X} = \begin{pmatrix} 0 & 0 & 1 \\ 0 & 0 & -2 \\ 0 & 0 & 1 \end{pmatrix}$$

$$\mathscr{Z} = (1/17) \begin{pmatrix} 10 & 12 & 14 \\ 0 & 0 & 0 \\ 5 & 6 & 7 \end{pmatrix}$$

$$\mathscr{C} = (1/17) \begin{pmatrix} 5 & 6 & 7 \\ -10 & -12 & -14 \\ 5 & 6 & 7 \end{pmatrix}$$

are constant matrices independent of the decurving parameters λ_i. They also form a closed set $\mathscr{S}$ under multiplication, as seen in the following multiplication table where $\mathscr{O}$ is the null matrix:

$$
\begin{array}{c|cccc}
 & \mathscr{I} & \mathscr{X} & \mathscr{Z} & \mathscr{C} \\
\hline
\mathscr{I} & \mathscr{I} & \mathscr{X} & \mathscr{Z} & \mathscr{C} \\
\mathscr{X} & \mathscr{X} & \mathscr{X} & \mathscr{C} & \mathscr{C} \\
\mathscr{Z} & \mathscr{Z} & \mathscr{O} & \mathscr{Z} & \mathscr{O} \\
\mathscr{C} & \mathscr{C} & \mathscr{O} & \mathscr{C} & \mathscr{O}
\end{array}
\qquad (5.4.3)
$$

$\mathscr{C}$ is a commutator, $\mathscr{C} = [\mathscr{X}, \mathscr{Z}] = [\mathscr{C}, \mathscr{Z}] = [\mathscr{X}, \mathscr{C}]$. In this table we read for instance $\mathscr{X} \cdot \mathscr{Z} = \mathscr{C}$ and $\mathscr{Z} \cdot \mathscr{X} = \mathscr{O}$. So, to the simple geometrical hierarchies generated by generalizing the decurving procedure, there corresponds an even simpler algebra of decurving matrices. From (5.4.3), one obtains the eigenvalues μ_i of $\Pi\Omega$ and their corresponding eigenvectors $\mathbf{V}_i(i = 1, 2, 3)$:

$$\mu_1 = 1, \quad \mu_2 = \prod \lambda, \quad \mu_3 = \prod v_i$$

$$\mathbf{V}_1 = \begin{pmatrix} 1 \\ -2 \\ 1 \end{pmatrix}, \quad \mathbf{V}_2 = \begin{pmatrix} 6 \\ -5 \\ 0 \end{pmatrix}, \quad \mathbf{V}_3 = \begin{pmatrix} 2(\prod v_i - 1) - \Delta \\ 2\Delta \\ (\prod v_i - 1) - \Delta \end{pmatrix} \qquad (5.4.4)$$

μ_3 is the Perron eigenvalue, $\mu_1 \ll \mu_2 \ll \mu_3$, for an infinite sequence of decurving operations. All eigenvalues are independent of the order of the decurving operations. Furthermore, the eigenvectors $\mathbf{V}_1$ and $\mathbf{V}_2$ are constant, independent of the number and intensity of decurving operations as well as of their sequence. But the information that they carry is not of immediate use, since any linear combination of $\mathbf{V}_1$ and $\mathbf{V}_2$ will have some negative components, corresponding to the negative number of sites of some coordination.

In the limit of an infinite sequence of decurving operations (Euclidean limit) the main structural characteristics are completely given by the Perron eigenvector $\mathbf{V}_3$. The structure of the completely decurved polytope depends only weakly on the order of the sequence of decurving parameters $\{\lambda_i\}$ through Δ. Moreover, because every $v_i \gg \lambda_i > 1$ (equation (5.4.2)), Δ can be approximated, quite accurately, by

$$\Delta \simeq (\lambda_w - 1) \prod v_i \tag{5.4.5}$$

where w labels the last decurving operation in the infinite series, so that the dependence on the sequence ordering is further reduced to the selection of the last operation in this series. Let us again emphasize that in practice a small number of decurving operations already yield enormous structures with very low curvature (for example, after two decurving operations Ω_3 and Ω_5 the structure contains 190 800 atoms).

Some quantities of physical interest are independent of the order of the decurving operations, like the following.

(i) The total number of atoms:

$$\frac{1}{120} N = n_{12}^{(w)} + n_{14}^{(w)} + n_{16}^{(w)} = 3\frac{5}{17}\left(\prod v - 1\right) \tag{5.4.6}$$

   or, approximately ($\lambda/v \ll 1$),

$$N \simeq 360\frac{5}{17}\,\text{Trace}\left(\prod \Omega\right) \tag{5.4.7}$$

(ii) The elastic energy which, to first order, is proportional to the total length of the disclination segments:

$$\frac{1}{120} E = n_{14}^{(w)} + 2n_{16}^{(w)} = 2\frac{5}{17}\left(\prod v - 1\right) \simeq 2\frac{5}{17}\,\text{Trace}\left(\prod \Omega\right) \tag{5.4.8}$$

   since $\lambda/v \ll 1$. These are the two independent relations between the three components of $\mathbf{V}_3$. One obtains immediately another important conserved quantity.

(iii) The average coordination number:

$$\bar{z} = (12n_{12} + 14n_{14} + 16n_{16})/N = 40/3 = 13.333 \tag{5.4.9}$$

$\prod \Omega_i$ generates hierarchical structures, characterized by $\mathbf{V}_3$ and parametrized by sequences of λ_i. It is interesting to recall in this context that all tetrahedrally close-packed phases (Frank–Kasper phases) have $13.3 < \bar{z} < 13.5$ (see table 7.1 and appendix A6). Spatial disorder, absent here where the transformations act globally, is essential to make $\bar{z}$ fluctuate (by introducing anisotropy or fluctuations in the size of coordination polyhedra (Rivier 1982a)).

5.4.3 Non-uniform decoration and spatial disorder

Up to now, we have supposed that, at a given stage of the decurving, all tetrahedral cells of the structure are identically decorated. But different decoration procedures, at different places in the structure, if they were compatible, would constitute a very efficient way to generate more disorder. It is clearly impossible to divide two neighbouring tetrahedra by two different decorations associated to different scaling factors λ_1 and λ_2. First neighbour distances will be different, which will prevent site matching on common edges and faces. Should one divide the common edge into λ_1 or λ_2 equal segments? But maybe it is possible to commute the order of two successive decurving operations at different points in space. The scaling is conserved if one tetrahedron is divided using $\Omega(\lambda_1)$ at the first step and $\Omega(\lambda_2)$ at the second step, while another nearby tetrahedron is divided by first using $\Omega(\lambda_2)$ and then $\Omega(\lambda_1)$. The two structures are locally different, because $\Omega(\lambda_1)$ and $\Omega(\lambda_2)$ do not commute, but the edge length is divided by $\lambda_1\lambda_2$ everywhere. An important property of this set of transformations is that the number of points generated by a sequence of operations $\Pi_i\Omega(\lambda_i)$ does not depend on the order of the operations $\Omega(\lambda_i)$: decorating a tetrahedron using $\Omega(\lambda_1).\Omega(\lambda_2)$ gives the same number of sites as decorating with $\Omega(\lambda_2).\Omega(\lambda_1)$. Furthermore, a study of the decoration of faces shows that the number of points on a face does not depend on the order of the decorations. If the two decorations yield the same geometrical decoration on a tetrahedron face, it would then be possible to mix $\Omega(\lambda_1).\Omega(\lambda_2)$ and $\Omega(\lambda_2).\Omega(\lambda_1)$ randomly in space, because interfaces would fit exactly even if local structures were different. This is not really the case, but a slight displacement of points at the interfaces allows a reasonable matching between the two tetrahedra without large structural change inside the cells and with an intermediate configuration at the interface. As far as the defect network is concerned, mixing randomly in space such pairs of transformations amounts to moving some disclination lines and creating a few closed disclination loops.

Two triangular faces of tetrahedra decorated respectively by using $\Omega(3).\Omega(5)$ and $\Omega(5).\Omega(3)$ are shown in figure 5.11, with the intermediate configuration, which allows a reasonable face matching. The number of points that have to be moved is small (12 out of the 90 points belonging to a face for $\Omega(3).\Omega(5)$).

A complete description of all new coordination polyhedra and new disclination segments appearing in these random structures is rather tedious (and not unique since there are two choices on the interface), but the main features can be summarized as follows.

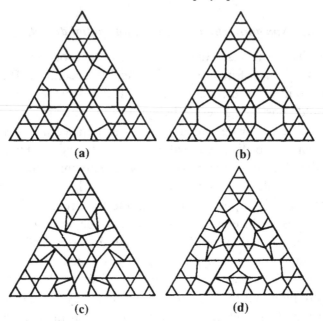

Fig. 5.11. Decoration of a tetrahedron face by two successive operations: (a) first $\lambda = 5$, then $\lambda = 3$; (b) first $\lambda = 3$, then $\lambda = 5$. Edges are divided into 15 segments. (c)–(d) Points obtained by the decoration are placed in a position intermediate between those of $\Omega(3).\Omega(5)$ and $\Omega(5).\Omega(3)$. These points are linked by edges according to $\Omega(3).\Omega(5)$ in (c), and to $\Omega(5).\Omega(3)$ in (d). The differences between the two configurations correspond to flipping a few edges.

(i) Some points with coordination number 15 and 13 appear.
(ii) Some edges are common to only four tetrahedra: this is a new type of positive disclinations (introducing disorder involves some recurving).
(iii) The disclination network generated by the iteration $\Omega(3)$ and the disclination network generated by the iteration $\Omega(5)$ are now connected by new disclination segments (positive or negative). These new segments form a 'pelota' (wool ball) close to the common interface separating two differently decorated tetrahedra.

In order to explain how flipping edges can generate such configurations of disclinations, we have described an example in appendix A6, §3, of what happens in a $\{3, 3, 5\}$ polytope when some edges are flipped.

Thus, using a non-uniform decoration, a new type of disorder is introduced, with new disclination segments. The density of defects increases, but these new defects can be called 'extrinsic', since their occurrence was not required in order to decurve the space, in contrast to the hierarchical networks generated by the uniform iterative decoration, which can therefore be called 'intrinsic'.

The energy cost of these new defects is not large (cf. §6.3); they may, in some cases, be entropically stabilized.

In the above example of $\Omega(3).\Omega(5)$ and $\Omega(5).\Omega(3)$ applied respectively on the two sides of a triangular interface, the coordination number changes are as follows (for each interface): 39 points change from 12 to 13 neighbours, 21 points change from 16 to 15 neighbours, 3 points change from 14 to 13 neighbours.

These new coordination polyhedra, which are nodes for positive and negative disclinations, are very anisotropic. This is known to increase the average coordination number (Rivier 1982a, Sadoc and Mosseri 1984) (see also in appendix A6, §2, and §5.4.2). To test this we can consider a $\{3, 3, 5\}$ polytope and apply $\Omega(3).\Omega(5)$ to 120 of the 600 tetrahedra and $\Omega(5).\Omega(3)$ to the remaining 480 tetrahedra. Taking account of the change of coordination numbers at the 480 interfaces, we find $z = 13.355$. In contrast, two steps of uniform decurving $\Omega(5).\Omega(3)$ yields $z = 13.320$. So, as expected, we find an increase in the coordination number together with the increase in defect density. In the limit of an infinite number of uniform iterations the asymptotic value is $z = 13.333\ldots$, whereas Coxeter (1958) derived a value $z = 13.39$ for a 'so-called' statistical honeycomb, which is closer to the more disordered structure. These non-intrinsic defects, which increase the disorder, are relevant for amorphous structures because they destroy the long range hierarchical orientational order of the polytopes and only keep some local order throughout the structure.

5.5 Deflation of the orthoscheme

5.5.1 The orthoscheme tetrahedron

The orthoscheme tetrahedron, described in detail in appendix A4, is a very efficient tool for describing a polytope. The faces of this tetrahedron are mirrors which can be combined together in order to generate all the symmetry operations of the polytope symmetry group. In this section, we describe the deflation–decoration procedure by looking at what happens inside an orthoscheme. We first take an example of the method in two dimensions.

5.5.2 A two-dimensional deflation: iterative flattening of an icosidodecahedron

The above iterative flattening method is much more easily understood in two dimensions than in three, and almost all of the ideas that we present now are generalized to one higher dimension. Note that, as has already been said above, the iterative flattening method can also be seen as an 'inflation', in the sense that the sphere containing the polyhedron (polytope) will have an inflated radius in order that the characteristic distance between vertices remains of the same order upon iteration.

So let us now apply the method to an icosidodecahedron. The icosido-decahedron (figure 5.12) is a semiregular (or Archimedean) polytope. It is denoted

$$\left\{ \begin{matrix} 3 \\ 5 \end{matrix} \right\}$$

(Coxeter 1973a, b) and shares the same symmetry group with the icosahedron $\{3, 5\}$ and the dodecahedron $\{5, 3\}$. Indeed it can be obtained by joining the mid-points of the $\{3, 5\}$ or $\{5, 3\}$ edges. Each vertex belongs to two regular pentagons and two equilateral triangles.

The icosahedral symmetry group has been described at length in appendixes A4 and A2. The full icosahedral group Y, including indirect transformations, is a group of order 120 which allows for a division of S^2 into a pattern of 120 spherical Möbius triangles, which are fundamental regions of this sphere tessellation (see figure A4.1). Any vertex configuration on S^2 having Y as a

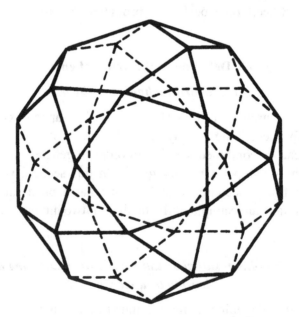

Fig. 5.12. The icosidodecahedron $\left\{ \begin{matrix} 3 \\ 5 \end{matrix} \right\}$.

symmetry group is completely defined by giving one fundamental region, and the distribution of vertices inside it. Therefore, the iterative flattening method will consist of an iterative decoration of a fundamental domain.

Note that the icosidodecahedron is found on the 'equatorial' sphere of polytope $\{3, 3, 5\}$. This choice is in fact governed by our aim to have a bridge between this two-dimensional example and the subsequent application of the iterative flattening method to the $\{3, 3, 5\}$ polytope. To an icosidodecahedron vertex, corresponds a point of coordinates (a_1, a_2, a_3) on the unit radius sphere S^2. The three orthoscheme vertices M_0, M_1, M_2 are vertices of the

$$\{3, 5\}, \quad \left\{ \begin{matrix} 3 \\ 5 \end{matrix} \right\} \quad \text{and} \quad \{5, 3\},$$

respectively, as can be verified by considering the orbits of these three points under successive reflections in the sides of the spherical triangle $M_0 M_1 M_2$.

We shall consider that the polyhedra under construction are either spherical, with bent faces and geodesic edges, or Euclidean, with flat faces and straight edges (chords). They are trivially related, the spherical polyhedron being the central projection of the Euclidean one onto the surface of the sphere S^2.

Let us now describe the method. In a first step the original polyhedron is constructed. It will be called hereafter the 'source' polyhedron P_0. It is characterized by the orthoscheme $M_0 M_1 M_2$ and by the location of a point M (or several points) inside it. P_0 vertices are the orbit of M under the Y^* symmetries. The second step consists of selecting a new triangle $M_0' M_1 M_2'$ which shares an angle with the orthoscheme $M_0 M_1 M_2$ (here the angle at vertex M_1, see figure 5.13). The triangle $M_0' M_1 M_2'$ is chosen so as to contain an integral number of orthoscheme replicas. Thus it contains several P_0 vertices. Note that the two spherical triangles $M_0 M_1 M_2$ and $M_0' M_1 M_2'$ do not have all their corresponding angles identical since they have different areas (recall that the area of a spherical triangle is proportional to the sum of its interior angles minus π).

The crucial point consists now of mapping the larger triangle onto the smaller one. In a flat space, this can easily be done with a homothety when the triangles are similar. In the curved space S^2, a homothety cannot be used because of the presence of an internal length scale (the radius of curvature). It is nevertheless possible to define homotheties along the geodesic lines $M_1 M_0'$ and $M_1 M_2'$ which make the points M_0' and M_0 and also M_2' and M_2 coincide.

But to ensure that all the points of the geodesic line $M_0' M_2'$ will lie on the geodesic line $M_0 M_2$ requires a continuous set of homotheties. This can be done

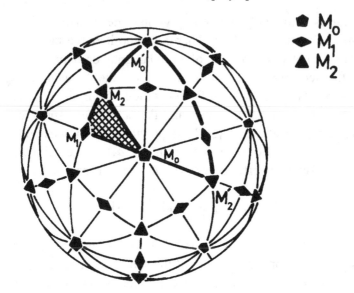

Fig. 5.13. The two spherical triangles $M_0 M_1 M_2$ and $M_0' M_1 M_2'$ which share the vertex M_1. Marks label vertices on S^2 where the symmetry is either five-fold, two-fold or three-fold. The 12 vertices (30, 20, respectively) which are the orbit of M_0 (M_1, M_2, respectively) are vertices of an icosahedron (icosidodecahedron, dodecahedron, respectively).

using barycentric coordinates. The first iterated polyhedron P_1 is obtained as the orbit under the group Y^* operations of the vertices lying in the orthoscheme $M_0 M_1 M_2$. It is represented in figure 5.14. The local order in P_1 presents similarities with that of P_0. Its description in terms of order and defects will be made in the next section. Let us just note that the configuration around the point M_1, is identical to what it was in P_0 (M_1 belongs to two pentagons and two triangles).

It is then easy to iterate the procedure again and build a new polyhedron P_2 by keeping the P_1 vertices lying inside M_0', M_1, M_2', identifying M_0, M_2 with M_0', M_2' and then getting the full P_2 as the orbit of the new points in the orthoscheme under the Y symmetries. As long as the two triangles keep specific relations (they have an angle in common and the big one contains an integral number of replicas of the first one), the successive polyhedra P_n will present very interesting properties. For example, as seen below, their defect set presents the same kind of regularity as the polyhedra themselves. From a topological point of view the asymptotic polyhedron P_∞ presents similarities to the Penrose tiling.

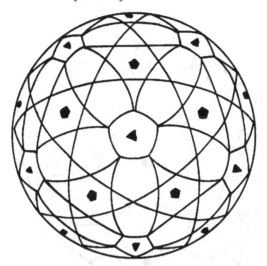

Fig. 5.14. The polyhedron P_1 and its defect set D_1. D_1 is the union of two sub-sets D_1' and D_1'' which are marked differently on the figure. A black q-gon inside a p-gon indicates that this p-gon is a disclinated q-gon.

5.5.3 The defect set

By definition, the source polyhedron P_0 is called the 'ordered (defect free) configuration'. It can be obtained by the local building rule: 'put two pentagons and two triangles around each vertex'. A definite value for the curvature of the underlying two-dimensional manifold is naturally associated with such a rule. This curvature takes a constant value at each site and is related to the so-called deficit angle δ_M at the vertex M:

$$\delta_M = 2\pi - \sum_{i=1}^{l} \theta_i^M \tag{5.5.1}$$

θ are the internal angles of the l flat polygons sharing site M. In two dimensions, owing to the existence of two famous relations, the Euler–Poincaré and the Gauss–Bonnet relations, simple and exact correspondence can be found between the geometry of the underlying manifold and the defect density (see chapter 4).

The iterative flattening method yields a very simple way to locate the defects and measure their intensity. Indeed a disclination point is generated at each vertex of the orthoscheme M_0, M_1, M_2 (and at all its replicas under the Y^* symmetries) whenever the angle is different from that of the corresponding triangle M_0', M_1', M_2'. Let us call D_p the defect set of a polyhedron P_p. D_p is

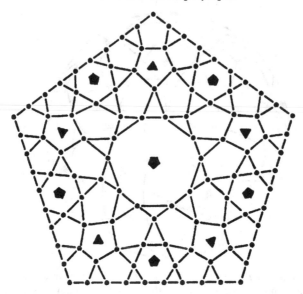

Fig. 5.15. Local view of the polyhedron P_2 and its local defect set D_2. The large pentagon which limits the figure is a pentagon of the original icosidodecahedron P_0.

the union of several sub-sets, each one being associated with a particular kind of disclination, and all sharing the symmetry group Y^*. For instance D_1 contains two sub-sets D_1' and D_1''. They are presented in figure 5.14; D_2 is presented in figure 5.15.

5.5.4 Three-dimensional case: iterative flattening of the polytope {3, 3, 5}

This method can be generalized to one higher dimension, here for the polytope {3, 3, 5}. We shall recover the above described hierarchical polytopes (with $\lambda = 3$). This is rather technical but nevertheless we present it briefly as it allows a better understanding of the complex and hierarchical defect structure, by analogy to the two-dimensional example.

The total symmetry group of polytope {3, 3, 5}, generates a regular division of S^3 into 14 400 spherical tetrahedra, a single one being called an orthoscheme (or fundamental tetrahedron). A tessellation is uniquely defined by the location of vertices inside an orthoscheme, the other vertices being generated by reflection in the faces of the orthoscheme.

Figure 5.16 represents one tetrahedral cell of polytope {3, 3, 5} and one orthoscheme inside it. The four vertices of the latter are located on one cell vertex, the centre of the cell, the centre of a face and of an edge. Each {3, 3, 5} tetrahedral cell contains 24 copies of the orthoscheme.

Once the orthoscheme is defined, it is very easy to generate any tessellation

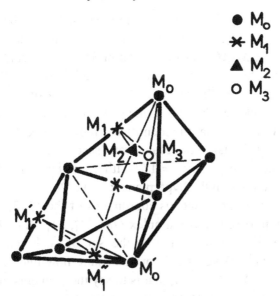

Fig. 5.16. On the top of the figure is a tetrahedral cell of the polytope $\{3, 3, 5\}$ with one particular orthoscheme inside it. The four vertices of an orthoscheme are a polytope vertex and an edge, face and cell centres. The coordinates of the 120 vertices of the $\{3, 3, 5\}$ are given by the 120 quaternions of Y'. One particular tetrahedral cell, with the north pole as apex, has coordinates:

$$(1, 0, 0, 0), (1/2)(\tau, 0, 1, \tau - 1), (1/2)(\tau, \tau - 1, 0, 1), (1/2)(\tau, 1, \tau - 1, 0)$$

where $\tau = (1 + \sqrt{5})/2$ is the golden ratio.

In the figure are the two basic tetrahedra of the iteration process. The figure shows four tetrahedral cells of the $\{3, 3, 5\}$ (heavy lines). The large tetrahedron M_0, M_0', M_1', M_1'' contains 20 orthoscheme replicas. Note that the index i in M_i or Mi' labels the sites according to the orbit in the symmetry group to which they belong. For instance M_1' and M_1'' are images of M_1 under given symmetry operations. They are both located at the middle of $\{3, 3, 5\}$ edges. The edges $M_0 M_1$, $M_1 M_3$ and $M_2 M_3$, belong to five-fold, two-fold and three-fold rotation axes, respectively.

which shares with the $\{3, 3, 5\}$ the same symmetry group G. It is enough to give the location of vertices inside the orthoscheme and then to apply the 14 400 symmetry operations. If the vertex has a generic position, one ends with 14 400 replicas (including the original vertex under the identity operation). If, however the vertex has a general position, its orbit may contain fewer points. For instance the image of $A = M_0$ gives rise to the 120 vertices of polytope $\{3, 3, 5\}$, while the image of M_3 gives rise to the 600 vertices of the dual polytope $\{5, 3, 3\}$.

Now we can begin the first iteration. We define a larger tetrahedron which shares the vertex M_0 with the orthoscheme and contains an integral number of orthoscheme replicas. Let us still consider the tetrahedron as being spherical. The three new vertices M_1', M_1'', M_0' belong to the same great circles as the geodesic edges M_0M_1, M_0M_2 and M_0M_3. Point M_0' is located at the $\{3, 3, 5\}$ vertex which belongs to a dodecahedral second neighbour shell surrounding the north pole. We have seen that the point M_3 is a vertex of the dual polytope $\{5, 3, 3\}$ which has dodecahedral cells, one of which also surrounds the north pole. Thus we can anticipate that the identification of M_3 and M_0' is equivalent to the identification of the large dodecahedron (second neighbour shell in the $\{3, 3, 5\}$) lying in the hyperplane $x_0 = 1/2$ with a smaller dodecahedral cell of the $\{5, 3, 3\}$ polytope lying in the hyperplane $x_0 = 0.926$.

Then we compute the homogeneous coordinates of the $\{3, 3, 5\}$ vertices in the barycentric system based on the simplex OM_0M_1', M_1'', M_0'. O is the centre of the hypersphere S^3. Let us call $\mathscr{M}_0$ the set of such points falling in OM_0M_1', M_1'', M_0'. Their barycentric coordinates are now interpreted as being based on the simplex $OM_0M_1M_2M_3$, that is to say the elements of $\mathscr{M}_0$ lie now in the small tetrahedron $M_0M_1M_2M_3$. But this tetrahedron is the fundamental region of the group G. It is then easy to generate a new polytope P_1 as the orbit under G of $\mathscr{M}_0$. More precisely $\mathscr{M}_0$ contains three points and the first iterated polytope P_1 has 2160 vertices. It is rather pedagogical to see where these 2160 come from. None of the three points in $\mathscr{M}_0$ is a generic point of the orthoscheme. Indeed, as said above, a generic point would have 14 400 replicas (the order of the group). Here, one point is located at M_0 and has 120 replicas. A second one is located somewhere on the edge M_0M_1 and has 1440 replicas. The third one is at M_3 and has 600 replicas (building a $\{5, 3, 3\}$), which gives the expected 2160 vertices of P_1. The local configurations around each vertex are not all identical (P_1 is not a regular polytope). However, strong orientational order remains in P_1.

It is now easy to iterate the procedure. We consider the new set of points of P_1 belonging to the large tetrahedron $M_0M_1'M_1''M_0'$. Using the same barycentric homothety as in the first iteration, the 14 elements of $\mathscr{M}_1$ are mapped into the fundamental region $M_0M_1M_2M_3$. A new polytope P_2 is then generated as the orbit of $\mathscr{M}_1$ under G, and contains 42 480 vertices. The iteration can proceed further on and larger and larger polytopes (in terms of their number of vertices) are obtained.

5.5.5 Description of the defect set

The formal definition of the defect set is very similar to the two-dimensional case, although its precise geometry is much more intricate. The initial polytope P_0 (the $\{3, 3, 5\}$) is the defect-free configuration. Defects (disclination lines) are generated along the edges of the fundamental region whenever the dihedral angle differs from that of the larger tetrahedron M_0M_1', M_1'', M_0'. Figure 5.16 represents the two tetrahedra $M_0M_1M_2M_3$ and M_0M_1', M_1'', M_0' and their relation with four tetrahedral cells of polytope $\{3, 3, 5\}$.

Let us call D_i the set of all the defect lines in polytope P_i. We have seen that D_1 consists of the edges of the $\{5, 3, 3\}$. The second iteration generates polytope P_2 with its associated defect set D_2. D_2 contains two disjoint parts as shown in figure 5.4.

6

Some physical properties

6.1 Structure factor of polytopes and orientational order

6.1.1 Structure factor in curved space

Scattering experiments on amorphous or liquid materials allow the determination of the structure factor, which depends only on the modulus of the wave vector. The structure factor, averaged over all orientations, is defined in Euclidean space by:

$$S(k) = \sum_{0}^{\infty} 4\pi r^2 \rho(r) \frac{\sin(k.r)}{k.r} \, dr \tag{6.1.1}$$

$\rho(r)$ is the number of atoms at the distance r from a reference atom taken as the origin and $4\pi r^2 \rho(r) \, dr = G(r) \, dr$ is the number of atoms in a spherical shell of thickness dr. This definition corresponds to an analysis in terms of spherical waves $\sin(k.r)/(k.r)$ where k is the modulus of the reciprocal space vector (Sadoc and Mosseri 1985b, Mosseri and Sadoc 1990). In a positively curved space, concentric spherical waves have an amplitude $\sin(l\omega)/(l.\sin\omega)$. As the spherical space is compact, there is only a discrete set of concentric waves: l is an integer. We note this number $n + 1$, by reference to the value of $\sin(k.r)/k.r$ for $k \to 0$ and the value of $\sin((n+1)\omega)/((n+1)\sin\omega)$ for $n = 0$, which are both equal to unity. In order to introduce a structure factor for the polytope we propose the following relations for the variables in spherical and Euclidean space:

$$4\pi \sin^2 \omega \sim 4\pi r^2$$

$$N(\omega) \sim G(r) \tag{6.1.2}$$

$$n + 1 \sim k$$

Table 6.1. *Values of ω_i and N_i in polytope {3, 3, 5} (see table A5.1)*

$\omega_i = 0$	$\pi/5$	$\pi/3$	$2\pi/5$	$\pi/2$	$3\pi/5$	$2\pi/3$	$4\pi/5$	π
$N_i = 1$	12	20	12	30	12	20	12	1

where we have supposed that the spherical space has unit radius.

We define the two atomic densities, $\rho(r)$ in Euclidean space and $\eta(\omega)$ in spherical space:

$$\rho(r) = \frac{G(r)}{4\pi r^2}$$

$$\eta(\omega) = \frac{N(\omega)}{4\pi \sin^2 \omega} \tag{6.1.3}$$

Now we can propose the following definition for the structure factor in spherical space:

$$S_n = \int_0^\pi 4\pi \sin^2 \omega . \eta(\omega) \frac{\sin[(n+1)\omega]}{(n+1)\sin \omega} d\omega \tag{6.1.4}$$

If we introduce a mean density η_0, it follows

$$S_n = \int_0^\pi 4\pi \sin^2 \omega [\eta(\omega) - \eta_0] \frac{\sin[(n+1)\omega]}{(n+1)\sin \omega} d\omega + t \tag{6.1.5}$$

where $t = 0$ for all $n \neq 0$, and $t = N_0$ for $n = 0$ (N_0 is the total number of vertices on the polytope). Therefore S_n is related to the Fourier component of the periodic function

$$\sin \omega \left(\frac{\eta(\omega)}{\eta_0} - 1 \right) \tag{6.1.6}$$

6.1.2 Structure factor of the {3, 3, 5} polytope and of hierarchical polytopes

In an ordered structure, like a polytope, the function $\eta(\omega)$ is a set of delta functions and consequently:

$$S_n = \frac{1}{n+1} \sum_i N_i \frac{\sin[(n+1)\omega]}{\sin \omega_i} \tag{6.1.7}$$

where N_i and ω_i are given in table 6.1 for the {3, 3, 5} polytope.

Owing to the symmetrical repartition of shells relative to the great sphere
($\omega = \pi/2$), only S_n with even n are different from zero. For $n < 60$, $S_n \neq 0$ for
$n = 0, 12, 20, 24, 30, 32, 36, 40, 42, 44, 48, 50, 52, 54, 56$. For $n > 60$, $S_n =$
$120 \times Int(n/60) + S_{n\,mod\,60}$ showing that $(n + 1)S_n$ is the sum of a periodic
function defined by $S_n = 120$ for $n = 0, 12, 20, \ldots, 56$ and a staircase function $120 \times Int(n/60)$. If we suppress the contribution at the origin, as is usually
done in conventional non-crystalline diffraction studies, a reduced structure
factor $s_n = (n + 1)(S_n - 1)$ can be defined. It is associated with the usual
function $i(k) = k.(I(k) - 1)$. The s_n function (figure 6.1) oscillates around zero
and can be compared to an amorphous metal structure factor.

Consider now a hierarchical structure obtained by an iterative process (like
those extensively described in chapter 5) which consists of replacing an atom
by a set of atoms with a well defined local configuration. A scaling factor λ
takes into account the ratio between the atomic diameter upon iteration. The
atomic motif is characterized by an interference function $I(k)$ which is similar
to a structure factor.

Suppose that the first iteration starts with the set of atoms defining the local
order, and that the second iteration replaces each atom by a similar set (with
the above scaling factor λ), then the interference function reads

$$J(k) = I(k) \times I(\lambda k) \tag{6.1.8}$$

If the iteration is repeated n times

$$J(k) = I(k) \times I(\lambda k) \times I(\lambda^2 k) \times \cdots \times I(\lambda^n k) \tag{6.1.9}$$

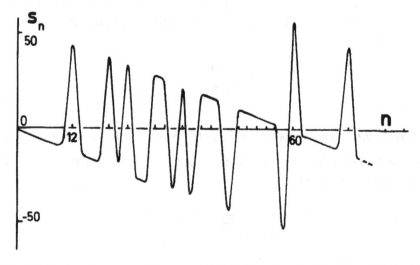

Fig. 6.1. Reduced structure factor s_n (slightly broadened) of polytope $\{3, 3, 5\}$.

This expression should give an approximate structure factor for the hierarchical polytopes. However, we have neglected orientation correlations between two neighbouring motifs. It is this effect which leads to interference spots in quasicrystals and in hierarchical structures. But in an amorphous structure it is reasonable to suppose that disorder screens this effect. Only a few terms contribute to $J(k)$ because for large $\lambda^n k$, $I(\lambda^n k)$ is close to 1. Figure 6.2 shows an interference function calculated this way from the polytope $\{3, 3, 5\}$, with one iteration and $\lambda = 3$ in the above expression.

Disorder in amorphous structures, modelled from hierarchical structures, will appear if different scaling factors λ are introduced, as discussed in chapter 5.

6.1.3 Orientational order

Nelson and Halperin (1979) and Nelson (1982) defined bond-orientational order by considering clusters of nearest neighbour 'bonds' surrounding every atom in two-dimensional structures. This parameter was used in characterizing the 'hexatic phases', intermediate between hexagonal crystals and liquids.

Steinhart *et al.* (1983) generalized the orientational order parameter in R^3. The case of curved structures was then analysed by Nelson and his coworkers. We summarize here the main lines of this approach (Sachdev and Nelson 1984). To quantify the orientational order embodied in a particular cluster of N sites, a density of bonds $\rho(\mathbf{r}, \Omega)$ is defined on a unit sphere surrounding a site at position $\mathbf{r}$, which can be expanded in spherical harmonics:

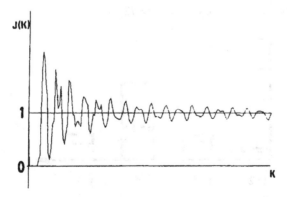

Fig. 6.2. Interference function obtained from polytope $\{3, 3, 5\}$ with one iteration and $\lambda = 3$ as explained in the text.

$$\rho(\mathbf{r}, \Omega) = \sum_{l=0}^{\infty} \sum_{m=-l}^{l} Q_{lm}(\mathbf{r}) Y_{lm}(\Omega) \tag{6.1.10}$$

In a fully isotropic liquid, all spatially averaged expansion coefficients

$$Q_{lm} = \frac{1}{N} \sum_{i=1}^{N} Q_{lm}(\mathbf{r}_i) \tag{6.1.11}$$

vanish, except the trivial term Q_{00}. For a bond cluster corresponding to the cubo-octahedral coordination polyhedron found in f.c.c. crystals, there are non-zero $\langle Q_{lm} \rangle$ characteristics of the cubic symmetry with $l = 4, 6, 8, \ldots$. Values of Q_{lm} with odd l vanish due to inversion symmetry. If the symmetry is icosahedral, on the other hand, the only non-trivial allowed spherical harmonics occur for $l = 6, 10, 12, \ldots$.

Steinhart *et al.* calculated the set of rotationally invariant bond-orientational order parameters

$$Q_l = \sqrt{\frac{4\pi}{2l+1} \sum_{m=-l}^{l} |Q_{lm}|^2} \tag{6.1.12}$$

The figure 6.3 shows their results obtained for a liquid simulated on a computer, with the classical Lennard-Jones potential

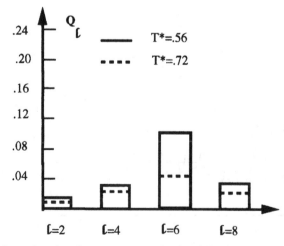

Fig. 6.3. The orientational order parameter obtained for two numerical simulations of a quenched liquid at different reduced temperatures (from Nelson *et al.* 1982).

$$V(r) = 4\epsilon \left[\left(\frac{\sigma}{r} \right)^{12} - \left(\frac{\sigma}{r} \right)^{6} \right] \tag{6.1.13}$$

They observed an increase in the value of Q_6 in terms of the reduced temperature kT/ϵ, upon cooling.

A complete study of the behaviour of the Q_{lm} indicates the presence of icosahedral order which grows with decreasing temperature. In R^3, quasicrystals are an important example of real structures presenting long range icosahedral order.

So, it seems interesting to define icosahedral order in S^3. The $\{3, 3, 5\}$ polytope is in this case the reference state which corresponds to the perfect icosahedral order.

6.1.4 Orientational order parameter in polytetrahedral structures

The Q_{lm} provide an orientational order parameter in R^3. Nelson and Widom (1984) went further in the analysis and introduced another order parameter which is in addition sensitive to a sort of translational order, of the kind found in polytope $\{3, 3, 5\}$. This allows us to formulate a Landau theory for melting on S^3, where the order parameters are not associated with reciprocal lattice vectors as in R^3, but adapted to the particular symmetries of the curved space.

The density $\rho(\mathbf{r})$ for an arbitrary configuration of particles on S^3 can be expanded in hyperspherical harmonics $Y_{n,m_1,m_2}(\mathbf{r})$,

$$\rho(\mathbf{r}) = \sum_{n=0}^{\infty} \sum_{m_1,m_2} Q_{n,m_1,m_2} Y_{n,m_1,m_2}(\mathbf{r}) \tag{6.1.14}$$

The coefficients Q_{n,m_1,m_2} in this expansion are like the Fourier coefficients in the classical equation for the density:

$$\rho(\mathbf{r}) = \sum_{k} \rho(\mathbf{k}) e^{i\mathbf{k}.\mathbf{r}} \tag{6.1.15}$$

The equivalent of the powder average structure factor $S(q)$ is

$$S_n = \frac{1}{(n+1)^2} \sum_{m_1,m_2} \langle |Q_{n,m_1,m_2}| \rangle^2 \tag{6.1.16}$$

In a perfect crystalline structure, the structure factor $S(k)$ is a set of delta peaks. In a perfect spherical structure, like the polytope $\{3, 3, 5\}$, only definite values of S_n are non-zero. The only allowed spherical harmonics for polytope $\{3, 3, 5\}$ occur for

$$n = 0, 12, 20, 24, 30, 32, 36, 40, 42, 48, 50, 52, 54, 56, 60,$$

and all even n. Note that we recover the same values, for the same S_n, as those given in the preceding section from an analysis in spherical waves.

In fact, the two approaches differ only in the correspondence between integers n, characterizing spherical waves in S^3 or hyperspherical harmonics, and k reciprocal vectors in R^3. In the spherical waves analysis the relation is $n + 1 \sim k$. In their approach, Nelson and Widom identify k and $[n(n+2)]^{1/2}$, by reference to the eigenvalues of the Laplacian ∇^2 in curved space:

$$-\nabla^2 Y_{n, m_1, m_2} = n(n+2) Y_{n, m_1, m_2} \tag{6.1.17}$$

and the comparison with the Euclidean case

$$-\nabla^2 e^{iK.r} = K^2 e^{iK.r} \tag{6.1.18}$$

Without going deeper into this description, note that such an analogy needs to go to the limit of an infinite radius for S^3 and to infinite values for n, while keeping constant the ratio of these two quantities; in that case $n + 1$ and $[n(n+2)]^{1/2}$ take the same asymptotic value, and the two descriptions converge.

A numerical simulation of the rapid quenching of 120 atoms on S^3 (of appropriate radius), interacting with a repulsive potential, showed that S_n has a strong peak for $n = 12$. This system 'crystallizes' very naturally in the polytope structure (Straley 1986).

6.2 Specific volume variation in disordered solids: a simple model

6.2.1 The close-packed sphere filling factor

We first consider the $\{3, 3, 5\}$ polytope, with spheres (balls) placed in close contact on each vertex. The packing fraction has already been calculated (§2.4.3) and found to be $p_0 = 0.774$. Let us now approximate the packing fraction resulting after mapping onto R^3. We first consider a part of S^3 (with radius r) limited by a sphere S^2 of radius r_d (Sadoc 1981b). The enclosed (three-dimensional) volume reads

$$v_d = 2\pi r^3 \left(\omega - \frac{1}{2} \sin^2 \omega \right) \tag{6.2.1}$$

where $r_d = r \sin \omega$. Upon a 'geodesic' mapping, S^2 is transformed into a Euclidean sphere of radius $R'_d = R\omega$ and volume

$$v'_d = \frac{4}{3} \pi r'^3_d \tag{6.2.2}$$

So, the volume expansion factor is v'_d/v_d. For the same reason, we must take into account an expansion of the spheres centred on each vertex. Upon mapping, the packing fraction is changed by the expansion factor:

$$c = \frac{(\omega - \frac{1}{2}\sin^2 \omega(\pi/10)^3}{(\pi/10 - \frac{1}{2}\sin \pi/5)\omega^3} \tag{6.2.3}$$

and $p = p_0 c$. As a parameter, ω is directly related to distortions: if ω is small, a small part of S^3 is mapped onto R^3 and distortions are small. The above calculated packing fraction is for finite clusters and not yet for an infinite space-filling structure. As an example, the value $p = 0.737$ is obtained for a single centred icosahedron, very close to the f.c.c. packing fraction. If space is filled by gluing replicas of a mapped polytope, cut surfaces are necessarily present (in the framework of the 'star-mapping' decurving process (Sadoc 1981a)). Near these surfaces, the density is lower than in other parts of the material and the packing fraction is therefore reduced. The cut surface area has been estimated and the best packing fraction can be calculated by optimizing with respect to two contradictory quantities: ω and the average surface area. The obtained packing fraction is $p \simeq 0.62$, very close to values measured in real-space ball packing or computer simulations.

We can also estimate the specific volume in the case, often discussed in this book, of a 'decurving' procedure leading to a network of disclination lines. Indeed, consider Voronoi cells of the decurved structure, assuming that the total area of the faces of a Voronoi cell around a given atom is proportional to the coordination number. The specific volume then reads

$$V_s \simeq \frac{n_{16}(16/12)^{3/2} + n_{14}(14/12)^{3/2} + n_{12}}{n_{12} + n_{16} + n_{14}} \tag{6.2.4}$$

in units of the volume of the Z_{12} cell. The fully decurved structure therefore has $V_s \simeq 1.18$ (instead of the curved space value $V_s = 1$), which corresponds to a variation, upon decurving, of

$$\Delta V_s \simeq \frac{0.54 n_{16} + 0.26 n_{14}}{n_{12} + n_{16} + n_{14}} \tag{6.2.5}$$

or more simply,

$$\Delta V_s \simeq \frac{\frac{1}{8}(4 n_{16} + 2 n_{14})}{n_{12} + n_{16} + n_{14}}$$

which can be generalized to any structure containing Z_i sites, under the condensed form (within the approximation $(1 + u)^{3/2} \simeq 1 + 3u/2$)

$$\Delta V_s \simeq \frac{\frac{1}{8}(\sum_i (i - 12) n_{14})}{\sum_i n_i} \qquad (6.2.6)$$

6.2.2 *Specific volume variation with temperature in simple metals*

Thermal variation of the specific volume of elements or simple compounds generally exhibits a small increase with T in the solid state, a discontinuity at the melting point and an important increase for the liquid specific volume at higher temperatures. Nevertheless, if the compound is in an amorphous state, the specific volume variation curve is continuous and can be divided into two regions: at low T, the specific volume is a few percent higher than in the crystalline state, the two curves being approximately parallel. At higher T, the glass specific volume curve is in continuity with the undercooled liquid one. Note that in many cases, as for liquid copper (Sadoc 1981b) for example, the extrapolated specific volume at 0 K takes a value smaller than in the crystalline structure (figure 6.4); this is related to the Kauzmann paradox (Kauzmann 1948).

A qualitative explanation can be given using topological arguments. The main hypothesis is that the extrapolation of the liquid structure to very low T

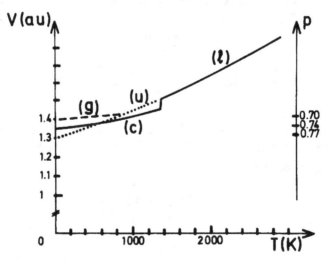

Fig. 6.4. Variation of the copper specific volume: (l) liquid, (c) crystal. Extrapolation to low temperature for the undercooled liquid (u). Estimated behaviour for glassy copper (g). The volume scale is that of a copper spherical atom and p is the packing fraction.

(the 'undercooled' (u) curve) leads to the ideal structure defined in curved space (it is indeed found that the extrapolated value is very close to the polytope value). In curved space we suppose that the number of defects increases with temperature, and that this is the source of the volume variation. At a given temperature T_0, the defect density is large enough to achieve a complete flattening in the Euclidean space. Above T_0, this hypothetical structure becomes physically observable and the usual liquid behaviour is recovered. The glass specific volume at $T = 0$ K is that of the polytope plus a contribution due to the intrinsic defects (those necessary to decurve the polytope). This is why the glass curve should increase slowly with T (a solid state behaviour) and reach the liquid curve at T_0.

6.3 Landau theory of frustrated systems

6.3.1 Basis of the Landau theory

We follow here the main lines of a pedagogical presentation given in a reference (Venkataramann *et al.* 1989). The Landau theory, sometimes qualified as a mesoscopic theory, is an approach which treats matter at a level intermediate between the atomic and the macroscopic level. Its basic principle consists of expanding the free energy with respect to an order parameter Ψ, under the form

$$F = F_0 + a\Psi^2 + b\Psi^3 + c\Psi^4 + \cdots \qquad (6.3.1)$$

The a, b, c coefficients depend on the temperature and pressure, etc. A more accurate treatment has to include inhomogeneities due to thermodynamic fluctuations. This is done using the 'Ginzburg–Landau' form for the free energy which contains a term proportional to the square gradient of the order parameter $|\nabla\Psi(r)|^2$.

It is convenient to expand the free energy using combinations of the order parameter invariant under G, the symmetry group of the most isotropic phase. Standard group theory is used in order to build these invariants. If we are interested by atomic structures, they are then characterized by a density function $\rho(r)$. A constant density ρ_0 is associated with the liquid phase, and, upon decreasing the temperature, the emerging order is described by a density $\rho(r) = \rho_0 + \delta\rho(r)$. The density modulation reads

$$\delta\rho = \sum_{i=1}^{n} \eta_i^{\Gamma} \phi_i^{\Gamma} \qquad (6.3.2)$$

where ϕ_i^{Γ} form a basis of functions which are transformed like the irreducible

n-dimensional representation of G. The coefficients η_i^Γ are the order para-meters which enter the theory. It is interesting to use normalized forms of these parameters, and so to write

$$\eta_i = \eta\gamma_i \quad \text{with} \quad \sum_i \gamma_i = 1$$

$$F(T, P, \eta) = F_0 + \eta^2 A(P, T) + \eta^3 \sum_\alpha B_\alpha(P, T) f_\alpha^{(3)}(\gamma_i)$$

$$+ \eta^4 \sum_\alpha C_\alpha(P, T) f_\alpha^{(4)}(\gamma_i) + \cdots \tag{6.3.3}$$

where $f_\alpha^{(3)}$ and $f_\alpha^{(4)}$ are the polynomial invariants (under then G action) formed from the γ_i.

In the case of a liquid–solid transition, the density modulation is expanded into plane waves indexed on a 'star' of reciprocal space vectors corresponding to the strong Bragg reflections of the ordered phase,

$$\delta\rho(r) = \sum_i \rho(k_i)\exp ik_i r$$

which gives

$$F = F_0 + A(k, P, T)\rho(k)\rho(-k)$$

$$+ B(k, P, T) \sum_{i,j,l} \rho(k_i)\rho(k_j)\rho(k_l)\delta(k_i + k_j + k_l)$$

$$+ C(k, P, T) \sum_{i,j,l,m} \rho(k_i)\rho(k_j)\rho(k_l)\rho(k_m)\delta(k_i + k_j + k_l + k_m)$$

$$+ \cdots \tag{6.3.4}$$

This form imposes, via the delta functions, the invariance of the free energy under translations. Using this kind of expansion, it has been possible to compare the different types of order which can appear from the liquid phase, as was first done by Alexander and McTague (1978) and then studied in detail in the field of quasicrystals (Venkataramann *et al.* 1989). One should also mention the work of Mercier and Levy (1983), done before the discovery of quasi-crystals, who were interested in icosahedral structures in the framework of the Landau theory.

6.3.2 Orientational order and curved space

The standard way to define order parameters suitable for the description of an orientational order is to replace the density $\rho(r)$ by a new density $\rho(r, \Omega)$ (see earlier sections in this chapter). The latter, constant in the liquid, should be sensitive to an emerging orientational order; it is expanded into the spherical harmonics (which here play the role that plane waves had in the case of translational order), with Q_{lm} as coefficients. Again using the group theory tools, for instance the $3 - j$ Wigner symbols, it is then possible to build invariants, under the $SO(3)$ group, on which the free energy is subsequently expanded.

In going one step further, Nelson and Widom have developed a Landau theory of dense polytetrahedral structures, using an order parameter related to the $\{3, 3, 5\}$ polytope. We have seen above that the first non-vanishing term in the expansion of the density of the $\{3, 3, 5\}$ polytope in S^3 into hyperspherical harmonics Y_{n,m_1,m_2} corresponds to $n = 12$. The chosen order parameter is therefore taken as Q_{12,m_1,m_2}. Notice that it is a 13×13 matrix (or a vector with 169 dimensions)! These authors then define a local order parameter $Q_{12}(r)$ in R^3 by placing at the point r a tangent sphere S^3 and then mapping, by inverse stereographic projection, the surroundings of r in R^3 onto S^3. But since the order parameter depends explicitly on r, the gradient term in the free energy must be included. One then faces an additional difficulty related to the spherical space used as a reference. It is not enough to define the order parameter at isolated points r in R^3, it is also necessary to have a rule in order to compare its values at neighbouring points (because of the gradient operator); in other words, we are here facing the fact that the system is frustrated. Along these lines, Sachdev and Nelson (1984) have derived a theoretical structure factor for metallic glasses which compares well to experimental results.

With the same general principles, Sethna has derived a continuum elastic theory, but with a simplest order parameter related to 4×4 matrices of $SO(4)$ (Sethna 1983). This theory can also be used in the case of frustrated blue phases, which also have an ideal structure in S^3.

6.4 Two-level systems

6.4.1 Hierarchical polytopes and two-level systems

An important property of the non-uniform decorations described in the previous chapter is that they generate two-level systems (TLS, or maybe better called 'tunnelling level systems'). As already mentioned, the distortion of the structure necessary to accommodate two different decorations is not unique. For instance, there is an ambiguity in the choice of the connection of points at interfaces separating decorations of the kind $\Omega(3).\Omega(5)$ or $\Omega(5).\Omega(3)$ (see figure 5.11).

A classical transformation between the two configurations requires cuts and reconstructions of bonds, but a tunnelling mode between two configurations can also be envisaged: this defines a TLS. Furthermore, it can be shown in the $\Omega(3).\Omega(5)$ example that, even if the interface configuration is fixed, the flow of points that are displaced in one tetrahedron has a rotational part relative to the axis orthogonal to the interface: so there are still two possibilities, and a small number (1 to 4, approximately 2) of TLS can be associated to an interface separating two configurations $\Omega(\lambda_1).\Omega(\lambda_2)$ and $\Omega(\lambda_2).\Omega(\lambda_1)$.

We therefore have in hand a microscopic model for TLS: let us try to learn something with it, by supposing on the one hand that the ideal order is well described by the polytope, and on the other hand that the source of disorder is this non-homogeneous decoration. In the above example, decorating one fifth of the 600 tetrahedra of the $\{3, 3, 5\}$ with $\Omega(5).\Omega(3)$ in place of $\Omega(3).\Omega(5)$, leads to $480 \times 2 = 960$ TLS in a structure containing 190 800 atoms. Note that this concentration of 0.005 TLS per atom is rather high compared with experiment: $(10^{-4}-10^{-5}$ (Black 1981)), which may indicate that real structures are far from being saturated with these defects. We can also estimate in this example the approximate number of atoms that are moving in every TLS, by looking at those points that have their coordination number changed: 63 atoms per TLS. It is also possible to define TLS involving a larger number of atoms. They will have a larger associated energy. Using the structures generated from the $\{3, 3, 5\}$ polytope with $\Omega(3).\Omega(5)$ in some places and $\Omega(5).\Omega(3)$ in other places, it is possible to derive new structures by doing uniform decorations at later iterations. This is always possible because the structure is still decomposable into tetrahedra. In the matrix formulation, one should then take into account new sites, of type Z'_{10}, Z'_{13} and Z'_{15} (see the discussion in appendix A6 on canonical Frank and Kasper sites). In fact, the number of Z'_{13} and Z'_{15} sites does not change upon subsequent iterations, and so only those sites introduced by the non-uniform operations remain (with an increase of their respective distance upon subsequent iterations). On the other hand, new Z'_{10} sites are appearing at each iteration, as intermediate points lying on the positive disclination segments introduced by the non-uniform operation. These sites contribute to the energy proportionally to the length of these lines.

Defects associated with the interfaces become more and more diluted in the structure as new uniform iterations are performed. They are still related to TLS, but involve a very large number of atoms: $\Pi_i v_i$, where v_i are Perron roots associated with the next iterations. The TLS associated with these 'extended' defects are related to collective displacement of atoms. The energy barrier is higher, the tunnelling rate longer, and the energy splitting between the tunnelling modes smaller than with less extended defects. By considering all the non-homogeneaous decorations which consist of local commutations in the order of decurving processes described by $\Pi_i \Omega(\lambda_i)$, one then gets a hierarchy of TLS, at all scales in the structure, which would lead to a broad density of TLS per energy range.

6.4.2 Elastic energy of homogeneous structures

We wish now to discuss the stability of these configurations and their relative energy. Let us call E_0 the ('ideal') cohesive energy per atom for the perfect polytope in curved space. By definition, it should correspond to an absolute minimum and will therefore be used as a reference. Suppose now that the additional energy of a decurved structure (an elastic strain energy) is simply proportional to the total length of the disclination network. We assume that all disclination edges carry the same energy (since they correspond to the same rotation angle), and that all nearest interatomic distances are equal. In this approximation, we obtain a decurving energy per atom which reads

$$\Delta E_g \text{ proportional to } \frac{4n_{16} + 2n_{14}}{n_{12} + n_{16} + n_{14}} \tag{6.4.1}$$

in the uniformly decurved structure. It is interesting to note that the above discussed specific volume follows a similar relation: ΔE_g and V_s are proportional, and the specific volume can be used to measure energy variations. Note that $(4n_{16} + 2n_{14})$ is independent of the order among the decurving operations (see chapter 5).

6.4.3 Interface energy for non-homogeneous decorations

Let us estimate this energy in the case $\Omega(3).\Omega(5)$. Under the assumption that positive and negative disclinations (with the same rotation angle) carry the same energy, we have

$$\Delta E'_g \simeq \frac{\frac{1}{8}(4n_{16} + 2n_{14} + 7n_{13} + 9n_{15})}{n_{12} + n_{16} + n_{14} + n_{13} + n_{15}} \tag{6.4.2}$$

A Z'_{13} site is described in appendix A6. It has one positive and two negative disclinations incident onto it. Nevertheless, it seems that in the present case a disclination quadrupole appears in the coordination shell of a Z'_{13} site. Then, seven disclination segments meet on this type of site. In addition, each Z'_{15} site is associated here with four negative and one positive disclination segments with possibly also a quadrupole. This estimation (seven for Z'_{13} and nine for Z'_{15}) for the number of incident segments is an upper bound, while the simpler estimate (three for Z'_{13} and five for Z'_{15}) is a lower bound. Note that these coordination shells are far from being canonical Frank and Kasper Z_n and are considerably distorted.

The difference in energy between homogeneous and non-homogeneous decurvings (ΔE_g and $\Delta E'_g$) can be expressed in terms of interface effects induced by the local commutation of decurving operations. With $\delta n_{14} = -3$, $\delta n_{16} = -21$, $\delta n_{12} = -39$, $\delta n_{13} = 42$ and $\delta n_{15} = 21$, the variation of the n_i introduced by an interface, we get

$$\Delta E_f = \frac{\frac{1}{8}(2\delta n_{14} + 4\delta n_{16} + 7\delta n_{13} + 9\delta n_{15})}{n_{12} + n_{16} + n_{14} + n_{13} + n_{15}} \tag{6.4.3}$$

We obtain $\Delta E_f = \frac{1}{8}(393/n_{tot})$ as the upper bound and with $\Delta E_f = \frac{1}{8}(141/n_{tot})$ for the lower bound.

For a non-homogeneous decurving process, the specific energy relative to the curved space energy reads

$$\Delta E'_g = \Delta E_g + F\Delta E_f$$

where F is the number of interfaces which are the locus of local commutations. Still using the specific volume as an energy unit, we have

$$\Delta E_g \simeq \frac{\frac{1}{8}(4n_{16} + 2n_{14})}{n_{12} + n_{16} + n_{14}} \tag{6.4.4}$$

or, following the numerical estimate $\Delta E_g \simeq 0.18$, and $\Delta E_f \simeq 2.5 \times 10^{-4}$ or 0.85×10^{-4}, depending upon which bound (upper or lower) is used for the Z'_{13} and Z'_{15} energy.

These numerical values are obtained in the example of a $\{3, 3, 5\}$ polytope decorated using $\Omega(3).\Omega(5)$, with some local commutations; in this case the number of vertices $n_{tot} = 190\,800$. This large model is a good approximation for any structure obtained with a series of iterations terminated by $\Omega(3).\Omega(5)$, and with some local commutations made only at these last two steps.

The energy of permutations occurring earlier in the decurving series is due to extra disclinations introduced by the commutation. When further decurving operations are performed, these extra disclinations have their lengths changed, and so the energy of interfaces created 'deeply' in the decurving series is:

$$\Delta E^* = \Delta E_f \prod_i \lambda_i \qquad (6.4.5)$$

where the λ_i are the length scales characterizing the later iterations. These energies can therefore be very large if the 'extended defects' of commutation have occurred at an early stage of the decurving series.

6.4.4 Back to melting and the glass transition

We have briefly described above how frustration effects have inspired pheno-menological models for undercooled liquids. Annealing of the previous ex-tended 'commutation defects' implies collective displacements over large distances ($\simeq \Pi_i \lambda_i$). It is then reasonable to suppose that if these defects can be generated by thermal activation, this means that the structure allowing for such changes on large scales is better described by a liquid state than by a solid state. Furthermore, we expect that these defects, associated with a large energy, appear mainly when less extended defects (corresponding to commutations at the end of the series of iterations), are saturated. So we can estimate a melting energy from the energy of creation of local defects

$$\Delta E_m \simeq \Delta E_g + F_{max}\Delta E_f \qquad (6.4.6)$$

where F_{max} is the maximum number of interfaces (normalized per atom) between regions of local commutation occurring at the last stage of decurving.

Using the example of a $\Omega(3).\Omega(5)$ decoration in the polytope $\{3, 3, 5\}$, we can estimate $F_{max} \simeq F_{tot}$, where F_{tot} is the number of faces in the polytope ($F_{tot} = 1200$). We have already seen that decorating one fifth of the polytope tetrahedra generates 480 interfaces.

We can now propose the following qualitative picture. The glass transition is defined according to the change of the specific volume, which is approximately constant below T_g (neglecting the asymmetry of the pair potential) and increases linearly above T_g. Below T_g, we suppose the structure to be almost perfect, that is containing only the intrinsic defect density required by the decurving to flat space, and having no commutation defects (or a few frozen such defects which slightly increase the specific volume). Above T_g, local commutation defects are first thermally activated, while above T_m, extended defects are now generated (see figure 6.5). If we consider that energy is proportional to the specific volume, we can relate T_m and T_g:

$$\frac{T_g}{\Delta E_g} = \frac{T_m}{\Delta E_g + F_{max}\Delta E_f} \qquad (6.4.7)$$

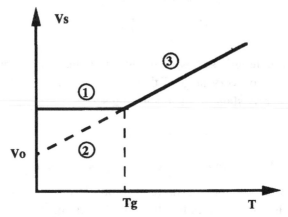

Fig. 6.5. Amorphous or liquid specific volume (neglecting thermal expansion). V_0 is the specific volume in curved space. (1) specific volume of a homogeneously 'de-curved' structure. (2) Continuous increase of intrinsic defect content during 'decurv-ing'. (3) Creation of extrinsic defects by thermal activation; these defects are related to the commutation.

Using the above estimated values, we get

$$1.3 T_g < T_m < 1.8 T_g$$

depending on the estimate of the Z'_{13} and Z'_{15} energies.

With this simple model, we therefore find a rather good agreement with experimental results, showing that in many glasses (Elliot 1983) melting temperature and glass transition temperature are empirically related by:

$$T_m \simeq 1.5 T_g$$

6.4.5 *Two-level-system energies*

Figure 6.6 gives a (very) schematic diagram of the energy landscape in configuration space, for a structure, containing a commutation defect at an interface, generated at the end of the series of iterations. ΔE_g is the energy needed to decurve the ideal polytope, which is related to an intrinsic density of disclinations (for instance a perfect hierarchical structure), while ΔE_f char-acterizes the interface defect, and therefore an additional disorder.

There are two kinds of energy barrier. One is between the homogeneous state without disorder (ΔE_g) and a state containing an interface defect, $\Delta E_g + F\Delta E_f$ (F is a small integer). The other corresponds to a flip between the different choices for the interface structure. It is this kind of barrier which is responsible for the TLS behaviour. These energy barriers can be estimated with

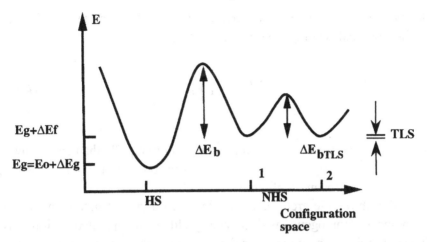

Fig. 6.6. Schematic energy landscape of an amorphous structure represented in the configuration space. E_0 is the energy in curved space and therefore the origin for all energy variations due to defects. Then $E_g = E_0 + \Delta E_g$ is the energy of the homogeneously decurved structure (HS) and $E_g + \Delta E_f$ is the energy of the two structures containing one commutation defect (NHS 1 and 2); $F = 1$, in fact it is a small integer. ΔE_b and ΔE_{bTLS} are the energy barriers between these configurations.

the $\Omega(3).\Omega(5)$ decoration in all cells except one decorated with a $\Omega(5).\Omega(3)$ operation. Two hundred and seventy sites are to be moved in order to create the commutated cell. In contrast, only 63 sites should move in order to flip the interface. The energies of these two configurations are very close, but we can estimate the barrier between them from the energy cost for moving atoms. Since only one elastic energy appears in the problem, it is reasonable to consider that the first barrier ΔE_b is close to ΔE_f, and that ΔE_{bTLS} only differs from ΔE_b by the number of sites implied in the displacement, which allows us to write:

$$\Delta E_b \simeq \Delta E_f, \quad \Delta E_{bTLS} \simeq \frac{63}{270} \Delta E_f \simeq \frac{1}{5} \Delta E_f \qquad (6.4.8)$$

The separation between tunnelling levels is related to ΔE_{bTLS}: the higher the barrier the smaller the separation.

The energy of more 'extended' defects, corresponding to earlier stages in the decurving process, reads

$$\frac{\Delta E_f^*}{\Delta E_f} \simeq \prod_i \lambda_i \qquad (6.4.9)$$

It scales with length, whereas the barrier ΔE_b^*, involving collective displacements of atoms, scales with the volume:

$$\frac{\Delta E_b^*}{\Delta E_b} \simeq \prod_i v_i \qquad (6.4.10)$$

Consequently, barrier heights increase much more rapidly than energy levels. Valleys, created at early stages of the decurving process, become very deep, and correspond to 'frozen' configurations.

Hence, the above hierarchical model leads to a rich structure in the configuration space. Homogeneous decurving yields well separated regions, with very close energies. Inhomogeneous decurving greatly complicates the energy landscape, multiplying valleys with a hierarchy of barrier heights. Thus, the kind of disorder (associated with this inhomogeneous decurving) should play an active role in a breaking ergodicity phenomenon. Note that this purely geometrical model displays the main features expected for two-level systems, the latter being introduced phenomenologically in order to describe both the glass transition and the low-temperature properties of glasses.

6.5 Frustration-limited domain theory

The concept of frustration is also at the basis of a recently introduced frustration-limited domain theory of supercooled liquid (Kivelsen *et al.* 1995). In this approach, the tendency of the molecules to pack into a locally preferred structure is frustrated by the postulated inability of such structures to tile space periodically. However, good glass-forming liquids are not monocomponent systems of sphere-like molecules, and a detailed description of the mechanism by which frustration operates is not available. It is assumed that frustration can be removed in some reference system in which a continuous transition to an ideal crystal would occur at a temperature T^*. In real space, frustration leads to a build-up of strain that resists the extension of the locally preferred structure and completely forbids the formation of the ideal crystal. When the temperature is lowered, the free energy loss due to the ordering process competes with the free energy increase due to the strain. At some point, not necessarily sharply defined, a balance is reached and the liquid breaks up into ordered domains whose size and further growth with decreasing temperature are limited by the frustration. Hence the name: theory of frustration-limited domains.

6.5.1 Supercooled liquids and the glass transition

Qualitatively, this picture allows us to rationalize the apparently anomalous behaviour of most glass-forming liquids. The stupendous continuous increase of structural relaxation times with decreasing temperature, that leads to the kinetic arrest at T_g known as the glass transition, is easily generated by considering the relaxation of a mesoscopic order parameter in frustration-limited domains. As is known from finite-size studies of ordinary systems below their critical temperature, relaxation of the order parameter indeed occurs on exponentially long and strongly temperature-dependent time scales even for modest size increase with T. Similarly, the heterogeneous and distinctly non-exponential character of the structural relaxation is naturally explained by the presence of a weakly polydisperse distribution of domain sizes. Another peculiar feature, namely the rapid increase of the entropy difference between the supercooled liquid and the crystal, can be understood as resulting from formation and growth of frustration-limited domains (G. Tarjus *et al.* 1998).

6.5.2 Scaling approach

At a more quantitative level, these phenomena can be described by means of a scaling approach based on a uniformly frustrated statistical mechanical model for a local order variable that is associated not with an individual molecule but with the locally preferred liquid structure. The effective Hamiltonian contains a reference part describing the continuous transition to the ideal crystal at T^* and uniform frustration is introduced via a weak but long-range competing interaction that generates the superextensive growth of strain opposed to the extension of the locally preferred structure. It is postulated, and checked in some model calculations (Chayes *et al.* 1991), that the critical point of the reference system at T^* is isolated and narrowly avoided because of frustration. This avoided critical point thus controls the physics of the whole system at lower temperatures and allows a scaling description about, but below, T^*. At still lower temperatures (possibly above and below T_g), the theory predicts the occurrence of a thermodynamic phase transition to what might be called a 'defect-ordered phase' by analogy with the Frank–Kasper phases of bi-metallic systems or the blue phases of liquid crystals. Although far from being unequivocally characterized, such a solid phase, distinct from both the crystal and the glass, may have been observed experimentally in glass-forming systems (Cohen *et al.* 1991).

6.6 Excitation spectrum

6.6.1 The case of crystals

The theory of the electronic structure of crystalline solids was developed very shortly after the introduction of modern quantum mechanics. Four years after the de Broglie thesis proposing a wave nature for the electron, and two years after the Schrödinger equation, Felix Bloch, in 1928, studied the propagation of an electron wave in a one-dimensional periodic potential, and showed how greatly the symmetry (periodicity) of the potential can simplify the calculation of the energy spectrum. In fact, Bloch was applying to the Schrödinger equation some properties already studied in the nineteenth century by the mathematician Floquet. In 1930, Léon Brillouin introduced the celebrated Brillouin zones, which mark out reciprocal space by lines and planes where the electron wave propagation is altered. One year later, the main difference between conductors and insulators was understood (by Wilson) in terms of the respective locations of Brillouin zones and the Fermi surface.

All these properties were studied in both extreme cases models, the nearly free electron, and the tight-binding model. In the latter case, one first considered idealized models of solids, like the linear chain and the two-dimensional square and three-dimensional cubic lattice. The simplest tight-binding Hamiltonian reads

$$H = t \sum_{\langle ij \rangle} |i\rangle \langle j| \qquad (6.6.1)$$

where t is the 'hopping' integral (which will be given the value one in the following) and the sum runs over all pairs of nearest neighbours. This simple 's' state Hamiltonian, which is *a priori* rather far from the real case, already displays some important aspects of the more sophisticated Hamiltonians.

The eigenvectors (Bloch waves) are associated with irreducible representations of the symmetry groups. A translationally invariant structure has a commutative group, whose irreducible representations are all one dimensional. The eigenspectrum is obtained through diagonalization, for each k value, of a matrix whose size n equals the number of sites in the unit cell. As a result, the eigenvectors are unequivocally labelled by the k vector defined in the first Brillouin zone, and a band index between 1 and n.

In amorphous systems, all the above simplifications are suppressed at once. The absence of periodicity forbids the use of the Bloch–Floquet theorem, the reciprocal space k vectors are no longer good quantum numbers and Brillouin zones can no longer be defined. As a result, electronic structure calculations are mostly purely numerical, the atomic structure being modelled by large clusters. One aim of the polytope models has been to describe amorphous systems in terms of ordered polytope-like regions interrupted by defects. We now see how it allows us to gain some new insights into the excitation spectrum. Indeed, we can now firstly calculate the spectrum of the ideal model in curved space and then analyse the effect of defect creation in the decurving process. We shall begin by looking at polyhedra in three dimensions, then calculate the excitation spectra of regular polytopes and finally investigate the role of disclinations. The main idea is to use, each time, Abelian sub-groups associated with symmetries which are believed to remain, at least locally and in an imperfect manner, after decurving.

6.6.2 Regular structures on S^1: excitation spectrum of polygons

Let us consider a periodic one-dimensional chain. Its spectrum is given by the standard dispersion formula $E(k) = 2\cos(ka)$ where k is a wave vector and a the intersite distance. Let us define now a supercell of size N and apply the Born–Von Karman (BVK) periodic boundary conditions. Geometrically, it amounts to closing the chain onto itself, and thus transforming it into an N-gon. No metrical considerations have to be taken since the above 'topological' tight-binding Hamiltonian is only dependent on the existence (or absence) of links between sites, and not on angles or distances. The net effect is to transform translational symmetry into a rotational symmetry. The BVK spectrum is obtained by computing $E(k)$ for the values $k = 2\pi n / Na$ with $n = 0, \ldots, N-1$. For example the spectrum of a square is obtained for $N = 4$, and one gets (with $t = 1$) the values $-2, 0, 2$, the middle one being doubly degenerate.

6.6.3 Regular structures on S^2: excitation spectrum of polyhedra

Consider now the periodic ladder shown in figure 6.7. The unit cell has two sites and the dispersion relation can be readily obtained: $E(k) = 2\cos(ka) \pm 1$. Note that for each value of k one has two solutions as expected from the doubly occupied unit cell. Taking a super cell of size $N = 4$, one gets a structure topologically equivalent to a cube. Computing the eigenvalues (at the four k values) gives $-3, -1, 1, 3$ with degeneracies $1, 3, 3, 1$ (figure 6.8).

A similar calculation can be done for the dodecahedron in the way depicted in figure 6.9, with $N = 5$ (Brodsky and DiVincenzo 1983, Mosseri *et al.* 1985).

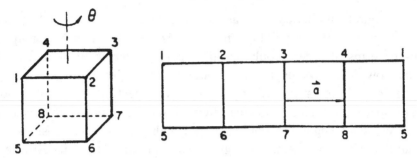

Fig. 6.7. The periodic ladder: a cube is unfolded around its four-fold axis. A $\pi/2$ rotation is related to a translation by a vector a. A similar analysis relates a pentagonal prism and a five square ladder.

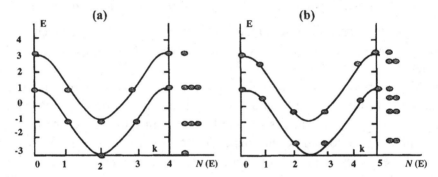

Fig. 6.8. (a) Energy levels for a cube in 's' bands in the Brillouin pseudo-zone related to the four-fold rotation. The density of state is obtained by mapping on the right axis. (b) In the case of a pentagonal prism, the dispersion curve is the same, but there are five k values.

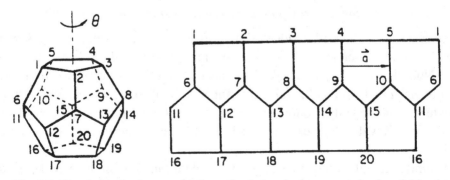

Fig. 6.9. The dodecahedron: unfolding of a dodecahedron around a five-fold axis.

6.6.4 Effect of a disclination on the excitation spectrum of a polyhedron

We focus on a disclination whose axis coincides with the rotation symmetry obtained upon BVK identification. Such a disclination transforms a cube into a pentagonal prism (with a $2\pi/4$ defect), and a dodecahedron into a 14-sided polyhedron, which is the dual of the Frank–Kasper Z_{14} (with a $2\pi/5$ defect). As far as the excitation spectrum is concerned, the dispersion relation remains unchanged, only N becomes $N + 1$. The dispersion relation forms an underlying constant skeleton common to the set of structures related by this kind of disclination.

On the spectrum presented in figure 6.8b, one can see that the defect produces a slight displacement and a slight broadening of the most regular spectrum (this is more explicitly discussed in the case of polytopes in the next section). This has to be compared, at least qualitatively, to a remark made by Baskaran (1986), see also §4.5.4, on the effect of Rivier lines on the electronic states. This effect has also been studied in continuous models, see for instance Furtado and Moraes (1994).

6.6.5 Regular structures on S^3: excitation spectrum of polytopes

The eigenspectrum of a hypercube can be calculated in a straightforward manner owing to its particular relation with a spherical torus. We have seen (appendix A3) that the 16 hypercube vertices are gathered on a 4×4 piece of a square lattice, wrapped along the two orthogonal directions. One can therefore use the dispersion relation of the infinite square lattice $E(k) = 2(\cos(k_x a) + \cos(k_y a))$ and solve it for k_x, $k_y = 2\pi n/4a$ with $n = 0, 1, 2, 3$. We get the eigenvalues $-4, -2, 0, 2, 4$, with degeneracies 1, 4, 6, 4, 1. Note the following point. The eigenspectra of the hypercube series (the square and the cube being the first terms) show the interesting feature that the eigenvalues are regularly spaced (spacing equal to 2) and the degeneracies follow the successive lines of a Pascal triangle. This can be easily proved and is left as an exercise for the interested reader. One clue is to proceed iteratively, using the eigenvectors of the $(n - 1)$-dimensional cube to construct those of the cube in dimension n. Recall that a given n-cube is built by taking two copies of $(n - 1)$-cube translated along the new direction.

For the $\{3, 3, 5\}$ spectrum, we have several possibilities according to which symmetry axis is selected. The highest symmetry order is the so-called '30/11' axis (figure 6.10) which is related to the simplicial helix (figure 2.7). Since the order of symmetry is 30, the initial 120×120 Hamiltonian matrix is reduced to a 4×4 matrix (Mosseri 1983, Mosseri et al. 1985) to be solved at the 30

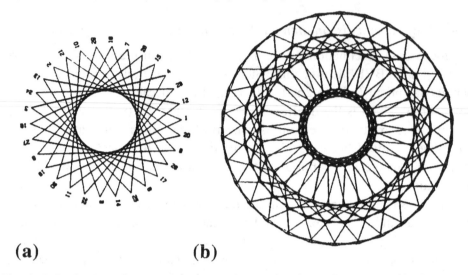

(a)   **(b)**

Fig. 6.10. Projection along a 30/11 symmetry axis of a polygon (a) and of a '240' polytope (b). In the case of the polygon, one can see that 11 turns are done when the 30 vertices are travelled through. The mapping of the $\{3, 3, 5\}$ polytope along the same axis is given in figure A5.2.

points of what is now called a 'pseudo-Brillouin zone' (figure 6.11). Notice that this spectrum can be fully solved using the full symmetry group of the polytope (Nelson and Widom 1984). But there is no simple way to extend this calculation to disclinated polytopes.

But this '30/11' symmetry axis is not the most interesting for introducing disclinations. Instead, the ten-fold screw axis is more promising because it corresponds to the disclinations which are introduced in chapter 4. It is even possible to go further in order to calculate, in the same way, the spectrum in the case where several disclinations are generated with the help of Hopf fibration (chapter 4) (Nicolis *et al.* 1986). These spectra (obtained with the hopping integral $t = -1$) are presented in figure 6.12 for the $\{3, 3, 5\}$ polytope, and polytopes D_{14}, D_{15}, D_{16}, D_{18} (with two disclinations and 168 sites, three disclinations and 180 sites, four disclinations and 192 sites, six disclinations and 270 sites), and also for the polytope P_1 obtained at the first step of the iterative construction (with 2160 sites) (see §4.1 and §5.1) (Nicolis *et al.* 1988). It is clear that new sites, and the lowering of the symmetry of the polytope, lead to a 'filling' of the density of states, which can be analysed as a broadening of the eigenlevels of the $\{3, 3, 5\}$ polytope.

Consider the most anti-bonding level in each spectrum. The unsymmetrical aspect of the state density is due to the well known effect of frustration.

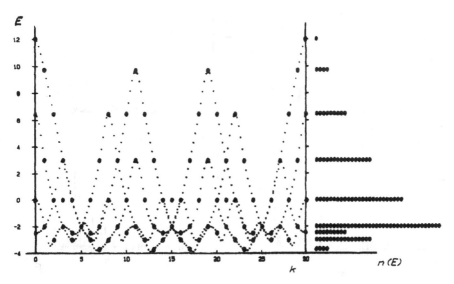

Fig. 6.11. The pseudo-Brillouin for the {3, 3, 5} polytope along the 30/11 axis, and its 's band' density of state.

Oddness of circuits in the structure prevents the building of completely anti-bonding states (which have wave functions with alternating sign on neighbouring sites). All dense structures contain a large fraction of triangles and are therefore sensitive to this effect. As an example, the spectrum in a f.c.c. structure runs from −12 to 4. We notice here that the anti-bonding edge for the {3, 3, 5} polytope is ≃ 3.71; this is simply due to the fact that the number of triangles (and tetrahedra) is larger in the polytope than in the f.c.c. structure. This remains for D_{14} and D_{15} polytopes, but disappears for the other presented examples of polytopes, even though their number of tetrahedra is large. This is probably due to the fact that the disclination procedure has changed a large number of five-fold rings into six-fold rings which decrease the frustration effect on the eigenstates. Note also that the P_1 polytope has a large density of states close to $E = +3$. This polytope has a rather low curvature, and is already a good model for tetrahedral close-packed structures.

It has been suggested that negative disclinations could attract electrons (Selinger and Nelson 1987), corresponding to a higher density of states close to disclinations. Calculating this explicitly for different polytopes, we have found that this depends on the filling of the band and is true only for low filling, with a reverse effect for high filling. So qualitatively we can say that, at low filling, electrons are more attracted by highly connected sites on which their wave

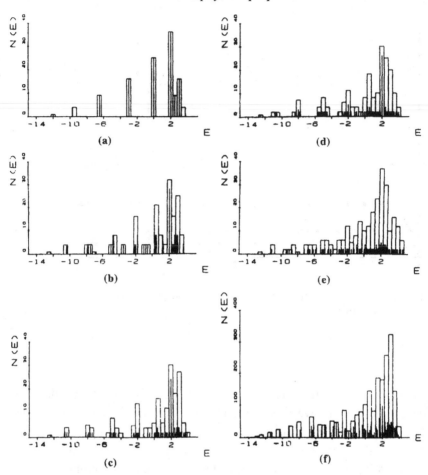

Fig. 6.12. The 's' band density of states of several polytopes: (a) $\{3, 3, 5\}$, (b) D_{14}, two disclinations and 168 sites, (c) D_{15}, three disclinations and 180 sites, (d) D_{16}, four disclinations and 192 sites, (e) D_{18}, six disclinations and 270 sites, (f) P_1 hierarchical polytope with 2160 sites containing a network of connected disclinations. Note that polytopes labelled by D_n contain only non-intersecting disclinations.

function can spread and so reduce the kinetic energy. At high filling, the same description applies for holes.

Another approach of the excitation spectrum of decurved structures has been proposed by Widom (1988), who uses the assumption that it is a statistical Coxeter honeycomb $\{3, 3, 5.104\}$ (see §4.3.3). The main idea is to evaluate the moments of the density of states, which are classically obtained by enumerating closed loops in the structure. The first four moments of this hypothetical structure have been evaluated.

6.6.6 Covalent structures on S³: the '240' polytope case

The '240' polytope is described in chapter 2, as an ideal model for covalent amorphous structure. Its excitation spectrum has been calculated, in the tight-binding approximation with a sp^3s^* Hamiltonian, using the 30/11 symmetry of the polytope (figure 6.13). Notice on the figure that the bands seem to repeat along a (pseudo) K axis with a periodicity $K = 2\pi/3$, a third of the fundamental periodicity $K = 2\pi$. This is due to the fact that 30/11 'channels' are built with three chains similar to those present in the diamond structure along the (1 1 0) direction; the symmetry operation T (related to K), returns to a given chain after three steps.

Optical properties are also interesting to look at. For a crystal, the translational symmetry imposes selection rules for optical transitions, $k_f - k_i = 0$. In amorphous structures it is usually supposed that the presence of disorder suppresses this selection rule. Effectively, experiments show a larger optical absorption in amorphous silicon than in its crystalline form (which has an indirect gap). Nevertheless, when the amorphous structure is related to an ideal curved space structure, the question of whether some traces of this perfection remain has to be asked. Indeed, there are symmetry-related selection rules in curved space for a polytope, and we can expect some weakened form of these

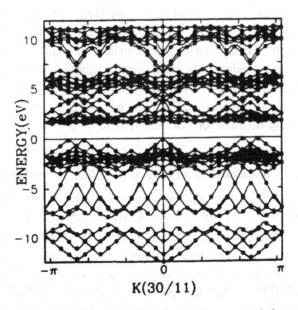

Fig. 6.13. Energy bands for the '240' polytope for a 'sp^3s^*' Hamiltonian. The eigenstates are indicated by black circles.

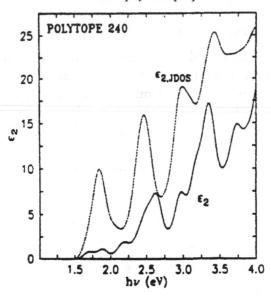

Fig. 6.14. The maginary part of the dielectric function plotted against the energy. The full curve takes into account selection rules, which are neglected in the dashed curve $\epsilon_{2,JDOS}$.

selection rules after mapping, when ordered regions still exist on length scales equal to the curved space radius.

Figure 6.14 shows ϵ_2, the imaginary part of the dielectric function plotted against the energy. The full curve has been calculated taking into account selection rules (with some drastic approximations), while the dashed curve $\epsilon_{2,JDOS}$ neglects them completely, being a simple convolution of the valence and conduction bands of the polytope (Brodsky and DiVincenzo 1983, Mosseri *et al.* 1985). Note that, mainly around the band edges, the remains of symmetries after mapping clearly decrease optical absorption.

6.6.7 *Vibration and melting of the {3, 3, 5} polytope*

It is also possible to consider the vibrational spectrum of polytopes (Widom 1986). Deng and Widom (1987) have in addition linked the effect of the vibrational properties of the {3, 3, 5} polytope to its stability, and have found that the latter is more difficult to melt than a f.c.c. structure. Using a simple repulsive interaction like $4/r^{12}$, they calculate a mean square displacement for atoms, $\langle u^2 \rangle = 0.0023 k_B T$ for the polytope and $\langle u^2 \rangle = 0.0027 k_B T$ for the crystalline structure. So at the same temperature, atoms oscillate more strongly in the crystal than in the polytope. The difference comes from the 'soft' acoustic modes which appear in the [1 1 0] direction of the crystal. The melting temperature of the polytope has been determined numerically (Straley 1986), which allows the deduction of the Lindemann constant, which measures the ratio between the mean square displacement of atoms at the melting temperature and the equilibrium distance between atoms. This constant is 0.19 for the polytope, compared to 0.15 in the crystal. The theoretical melting temperature of the polytope is nearly twice that of the crystal.

7

Periodic structures with large cells

7.1 Frustration and large cell crystals

We have described the occurrence of structural complexity in non-periodic systems as the result of geometrical frustration. This applies naturally to non-periodic structures like amorphous structures or quasicrystals, but also to some crystalline structures. In particular, it is interesting and instructive to analyse those periodic structures which also contain ingredients of frustration as a result, for instance, of local icosahedral configurations. We shall mainly focus on metals, or on clathrate structures which are dual from a geometrical point of view. Note that similar considerations could prove useful in systems with non-metallic interactions, for instance the C_{60} molecular crystal or the boron structure.

A major result of the above curved space approach is to propose a description of structural complexity at a range which is larger than that defined by the frustrated local interactions. Once the ideal structure has been defined in curved space, the appropriate range is related to the radius of curvature of the latter, which governs the average distance between the disclination decurving defects. The complexity is then encoded in the defect network itself. If ever this network adopts a periodic structure, this will often give rise to a large cell crystal. Since the paradigmatic example of frustration is to be found in close-packed metallic systems, it is then natural to look for their possible occurrence in nature. We shall also look at tetracoordinated (§7.4) and liquid crystalline (§7.5) periodic structures, in which similar networks can be recognized.

7.2 Complex structures in metals

In this section we present some examples of tetrahedral close-packed structures. First there are classical examples of Frank and Kasper phases (§7.2.1 and

7.2.2), then structures of pure metals (§7.2.3 and 7.2.4) and complex structures for simple alloys (§7.2.5 and 7.2.6). Sections 7.2.3 to 7.2.6 contain some difficult descriptions, because a complex structure is necessarily complex to understand in detail, but a rapid survey of figures presenting disclination networks and an inspection of table 7.1 should be sufficient for a general understanding of the mechanism of the complexity.

7.2.1 Frank and Kasper phases

Metallic systems are currently found with b.c.c. or f.c.c. structures, with two or four atoms per unit cell, but there exist several cases where a larger number of atoms per unit cell is found, the latter looking at first sight very intricate. Among these, Frank–Kasper or tetrahedrally close-packed phases have been the subject of renewed interest, because these crystalline phases are in some respect very close to quasicrystals (Janot 1992, Jaric 1989), and several new phases have been discovered recently. Since these structures keep the local tetrahedrally close-packed arrangement, it is very interesting to see how Nature accommodates the natural tendency of forming icosahedron-like configurations and the requirement of periodicity. We shall see now that these structures can be described by periodic networks of disclinations in a medium having the local order of the {3, 3, 5} polytope. Note that, in their original work, Frank and Kasper (1958) pointed out that the sites whose environment is not icosahedral form a network, with no end points, which they called the 'major skeleton', and which is now identified as the disclination locus. Appendix A6 is devoted to the different coordination shells found in Frank and Kasper phases and to their possible deviations.

An inspection of the Frank and Kasper unit cell geometries leads to the two following remarks:

- many atoms have icosahedral coordination shells (Z_{12} sites) with only slight deviations from the perfectness of the icosahedron;
- a finite proportion of atoms have a higher coordination (usually of type Z_{14}, Z_{15} and Z_{16}, the F.K. canonical cells) which form uninterrupted networks connected along the directions where the five-fold icosahedral symmetry is replaced by a six-fold local symmetry.

An immediate result is that the average coordination number is higher than 12, and comparable to that of the hierarchical polytopes and of the Coxeter statistical honeycomb (appendix A4, §2.2).

Note that in our process of decurving the polytope {3, 3, 5}, we were induced, upon introducing the disclinations, to generate precisely the same higher coordinated sites. It is then a natural tendency to identify the Frank–Kasper major skeleton with our disclination networks. But, to do so, one should first verify that the part of the unit cells complementary to these lines can be mapped, with some distortion, onto pieces of the $\{3, 3, 5\}$. Some remarks along these lines were made in 1982 (Sadoc and Mosseri 1982b), and the analysis done the year after (Sadoc 1983, Nelson 1983).

In the Frank and Kasper phases, the tetrahedra are (slightly) distorted. The ratio of the longest to the shortest edge is usually smaller than $4/3$ or even $5/4$ in Frank and Kasper structures but can be higher in some cases, as stated by Schoemaker and Schoemaker (1989). Note that the Frank and Kasper phases being mostly alloys, we shall meet mainly the largest atoms on the disclination lines, where the sites have higher coordination and therefore a larger atomic volume available.

7.2.2 Classical examples of Frank and Kasper phases

We first present the two classical examples of Frank and Kasper phases: the $A15$ structure, whose simplest example is the so-called β phase of tungsten, and the cubic Friauf–Laves phase. They show how a local tetrahedral arrangement can lead to periodic structures while keeping some icosahedral configurations.

The elementary cell of the $A15$ structure is a cube containing eight atoms. Inside the cube is a centred icosahedron, with the 12 outer atoms on the cube faces, which therefore count for $1/2$. The eight cubic corners are also occupied (counting for $1/8$). The three orthogonal two-fold axes of the icosahedron coincide with the three four-fold axes of the cube. This unit cell is represented in figure 7.1. Among the eight atoms of the elementary cell, two have an icosahedral coordination (Z_{12} sites or W_b) and six have a coordination polyhedron with 14 vertices (Z_{14} sites or W_a), the latter lying on disclination lines. So, we can say that the β phase structure is a $\{3, 3, 5\}$ polytope (with atoms on vertices), in which the disclination lines needed to flatten the curved space have changed the coordination polyhedron of three atoms out of four. This structure can be given the schematic formula $W_{a3}W_b$ and is also found in several alloys with the same stoichiometry, for instance Nb_3Ge, the well known superconductor. Nevertheless it is clear that this phase is not a stable phase of bulk pure tungsten: it seems to appear only with some oxygen in it (Weerasekera *et al.* 1994).

It is not obvious how to devise the precise process by which disclinations can be endlessly introduced, one by one, in the $\{3, 3, 5\}$ polytope, in order to get the $A15$ structure. We saw earlier (chapter 5), how this can be done for a

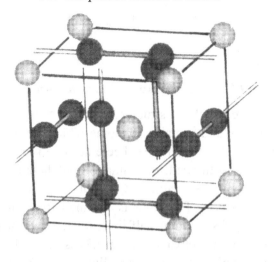

Fig. 7.1. The cubic unit cell of the β-W structure.

limited number of lines symmetrically disposed in the polytope, giving rise to a less curved polytope. Complete decurving was only obtained as the asymptotic result of iterative processes leading to either geodesic hyperdomes or hierarchical polytopes, the detailed, step by step, transformation being then given. Here, after having recognized the disclination network in the $A15$ crystal, we shall simply consider that it has been obtained in one step.

Friauf–Laves phases, for instance Cu_2Mg or C_{15} (Sadoc and Rivier 1987) have a cubic cell containing eight atoms of one type (Mg), with a large Goldsmith radius, arranged like carbon atoms in a diamond structure, and 16 atoms of a second type (Cu), with a smaller radius, filling the free space in the voids of the diamond structure. The coordination polyhedron for the latter type of site is a slightly distorted icosahedron with 12 vertices (it is a Z_{12} site), while, for the former type, it has 16 vertices which can be obtained from an icosahedron by inserting four half disclinations with tetrahedral symmetry, this site being therefore a Z_{16} site.

The structure of this cubic Laves phase can also be described as a stacking of sheets made of Friauf–Laves polyhedra and small tetrahedra, as represented in figure 5.7. All Cu atoms are on the vertices of these polyhedra, and Mg atoms are in the centre of the Friauf–Laves polyhedra. The Laves phase structure can be understood as originating from a $\{3, 3, 5\}$ polytope flattened by a network of disclination lines organized like the bonds of a diamond structure as shown in figure 7.2. It can also be viewed as an extreme case of the iterative procedure detailed in chapter 5, with $\lambda = \infty$. The mean coordination number (see table 7.1) is $\bar{z} = 13.33$ for the Laves phase and $\bar{z} = 13.50$ for the β-W structure.

7.2.3 *A pure metal Frank and Kasper phase: the β-uranium structure*

In this example of a pure metal, the structure is known to be very complex. The uranium cell is tetragonal and contains 30 atoms. There are five different types of site in this structure (Donohue 1974). There are two types of Z_{12} occupied by two and eight atoms in the cell. Coordination polyhedra of the β-uranium are like those in the β-W structure (Z_{14}), but there is also a 15-vertex polyhedron (see figure A6.1). The latter is obtained from an icosahedron by inserting three coplanar disclination half-lines intersecting at the centre (Z_{15} sites). The β-U disclination network (figure 7.3) is formed by two types of disclination line. One family consists of parallel lines going through Z_{14} sites (eight such sites in the cell). Disclinations of the second family remain in parallel planes and form a three-coordinated hexagonal network with nodes on

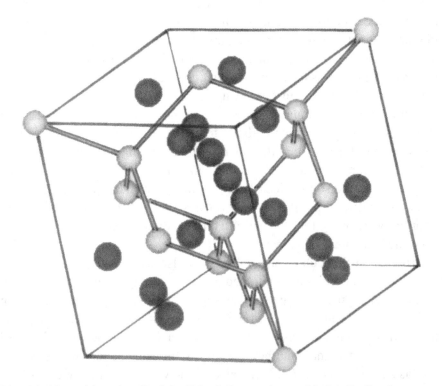

Fig. 7.2. The cubic unit cell of the Friauf–Laves phase with the disclination network.

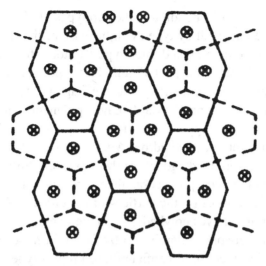

Fig. 7.3. The disclination network in β-uranium. There are two types of alternate planes containing disclination lines. The network of disclinations is drawn in full lines for one plane and in dashed lines for the next plane. There are also disclination lines orthogonal to the family of planes. They are represented by crossed circles.

the Z_{15} sites (four in the cell). Hexagons are not regular, with small edges corresponding to first neighbour distances of the structure, and larger edges going through intermediate Z_{14} sites (eight in the cell). The mean coordination number (see table 7.1) is intermediate between those of the β-W and the Laves phase structure, $\bar{z} = 13.46$.

7.2.4 A tetrahedral close packing: the α-manganese structure

This is another example of a pure metal with an extremely complex structure. There are 58 manganese atoms in a big b.c.c. cell ($I\bar{4}3m$) (Oberteuffer and Ibers 1970, Vieland and Wicklund 1974). There are four types of atomic positions in the cell which are, as usual, denoted I, II, III and IV. Just counting the coordination numbers, it is found that there are two type I (Z_{16}), eight type II (Z_{16}), 24 type III (Z'_{14}) and 24 type IV (Z_{12}) as presented in appendix A6. The type III atoms are usually given with a coordination number of 13, but in order to have a coherent tetrahedrization of the structure it is better to assign to them additional neighbours of type III atom. Its coordination polyhedron is not a canonical Frank and Kasper polyhedron because there are on this site three negative disclinations and one positive disclination.

The structure can be characterized by a packing of Friauf–Laves polyhedra of two types (I and II) corresponding to the two types of Z_{16}. The 12 vertices of these polyhedra are occupied by type IV in the case of the type I Friauf–Laves polyhedra and by nine type IV and three type III atoms for the type II Friauf–Laves polyhedra. There is a type I Friauf–Laves polyhedron centred on the centre of the cubic cell. Joined by their hexagonal faces in an eclipsed geometry, there are four type II Friauf–Laves polyhedra close to the central one. There are four other type II Friauf–Laves polyhedra sharing triangular faces with the same central polyhedron. All the centres of these Friauf–Laves polyhedra are aligned on the diagonals of the cubic cell. Sharing faces with the eight type II Friauf–Laves polyhedra, there are eight type I Friauf–Laves polyhedra with their centres on the vertices of the cubic cell. All atoms of the structure are on Friauf–Laves polyhedra or on their centre.

The disclination network is dominated by four-fold nodes corresponding to the Z_{16} sites, always in eclipsed configuration. From a type I site the network begins like a Z_{16} site of the P_1 polytope (chapter 4) obtained by one $\lambda = 3$ iteration on the $\{3, 3, 5\}$, but this ends on a type III atom. Then in order to connect a disclination network starting on the centre of the cell with one of the others starting on its vertices, there is an intricate 'pelota' of disclination segments. This 'pelota' connects four type III atoms defining a tetrahedral interstice of the structure. This (flattened) tetrahedron has its centre on the centre of a cell face (or a cell edge). Two orthogonal edges of the tetrahedron are positive disclinations, the four other edges being negative disclinations. The occurrence of this 'pelota' connecting Z'_{14} sites can be understood in terms of a conservation law imposed by Bianchi-like identities (Chapter 4).

Why a pure metal crystallizes in so intricate a structure is still an open question which is related to subtle electronic properties. The occurrence of more or less perfect icosahedral coordinations (type IV), favouring high local compactness is a partial answer, but it can be argued that there are simpler structures which offer this property. Why does Nature explore all possible structures?

7.2.5 Bergman and S. M. structures: hierarchy and tetrahedral close-packed structures

The $Mg_{32}(Zn, Al)_{49}$ crystalline alloy has a large cubic cell with 162 atoms and edge length 14.16 Å, known as the Bergman structure (Bergman *et al.* 1957), or the T-phase. The great interest in it comes from its relation to approximations of quasicrystals of close composition. We have described another, closely related, hypothetical structure, hereafter referred to as S. M., which has the same composition, but is slightly different. Both Bergman and S. M. structures are tetrahedral close packings, with the same sequence of concentric polyhedra around the vertices and the body centre of the cubic cell. The only difference is in the relative orientation of these clusters (figures 7.4 and 7.5).

The sequence of concentric polyhedra for both structures is: a central vertex (Z_{12} site), an icosahedron of Z_{12} sites, a dodecahedron of Z_{16} sites, an icosahedron of Z_{12} sites, a 'soccer ball'-like polyhedron (truncated icosahedron) and a truncated octahedron. Up to the third shell, remarkably enough, this shelling is identical to that surrounding a $\{3, 3, 5\}$ vertex (see table A5.1).

So, we are very close to what we have been looking for: a periodic large cell crystal with large parts modelled on the $\{3, 3, 5\}$ order. We must now see how these large clusters interconnect and what the disclination network looks like.

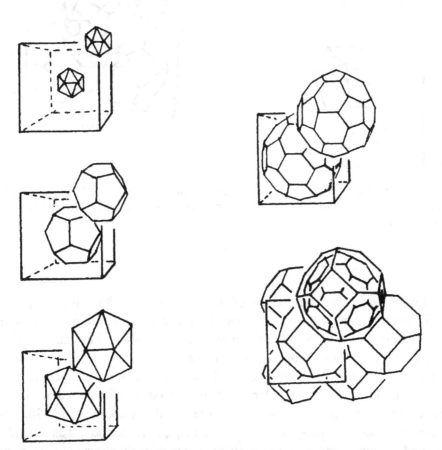

Fig. 7.4. The shelling in the body centred cubic Bergman structures: an icosahedron of Z_{12} sites, a dodecahedron of Z_{16} sites, an icosahedron of Z_{12} sites, a truncated icosahedron and a truncated octahedron.

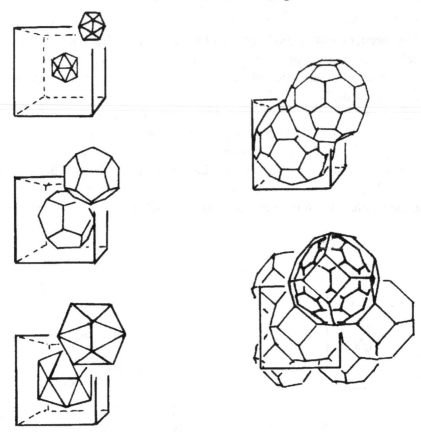

Fig. 7.5. The shelling in the modified Bergman structures, called S. M.: shells are the same as in figure 7.4, but with a different orientation around the cube centres and the cube vertices.

We already noted that the dodecahedral second shell is formed by Z_{16} sites, and is therefore part of the defect network. In order to proceed on the description of Bergman and S. M. structures, we now show how they can be interpreted as a decoration of the $A15$ structure.

In chapter 5 we have presented a decurving operation characterized by transfer matrices $\Omega(\lambda)$. But, interpreted as decoration of tetrahedra, nothing prevents us from applying it directly to a tetrahedral packing in R^3. Since no perfect tetrahedral packing can fill R^3, any t.c.p., even periodic, has an underlying undulating geometry, with positively and negatively curved regions corresponding locally to the number of tetrahedra shared by the edges (Regge-like approximation, see §4.3.4). Applying the iterative procedure directly on a t.c.p.

in R^3 will therefore produce a new t.c.p. with a new scale of undulations (related to the value of λ) in the underlying geometry.

Recall that, for $\lambda = 3$, a tetrahedron is decomposed into four tetrahedra and one centred Friauf–Laves (F.L.) polyhedron. Each hexagonal face separates two F.L. polyhedra, and constitutes with their two centres six tetrahedra sharing the edge joining their two centres, such an edge being therefore a segment of negative disclination.

The S. M. structure is simply obtained from an $A15$ structure on which the $\Omega(3)$ decurving operation is applied once. Getting the Bergman structure is more complex and we have now to introduce a more indirect method.

The $A15$ structure bears some relation to the body centred cubic structure: it contains icosahedra whose centres are on all the vertices of a body centred cubic lattice. But it is not a b.c.c. structure, since the icosahedron centred at the cubic cell body centre is rotated relative to the icosahedron centred at a cube vertex. Nevertheless, in order to place the icosahedra, it is worth using the body centred cubic skeleton and considering its dual Voronoi packing (sometimes called a Delaunay–Dirichlet decomposition). It is a packing of the so-called Kelvin polyhedra, which are semi-regular truncated octahedra made of squares and hexagons (see figure 7.6). It is a simple example of a perfect space filling by a single type of semi-regular polyhedron, different from a cube.

We first generate a structure, topologically identical to $A15$, by keeping consistently half of the Kelvin polyhedra vertices, plus all the original b.c.c. lattice sites. This amounts, on each square face, to keeping only two opposite

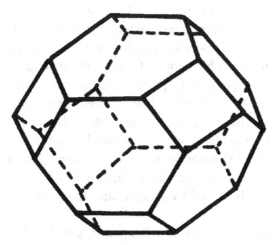

Fig. 7.6. Kelvin polyhedra are the Delaunay–Dirichlet domains of the b.c.c. structure.

sites. It is possible to make a coherent choice such as to align everywhere the square diagonals between the kept points with an $A15$ line skeleton (of half cell edge length).

However, this description is only exact in a topological sense, since the icosahedron obtained by selecting 12 out of the 24 Kelvin polyhedron vertices is distorted as compared to that found in β-W. So a metric distortion is needed in order to recover the exact structure: it amounts to changing square faces into rhombi. For simplicity, and also in order to make explicit the relation between Bergman and S. M. structures we shall limit ourselves to this topological relationship.

Square face diagonals can either be made to run through the whole structure as straight lines, which are indeed the disclination lines or major skeleton of the $A15$, or as 'antialigned' segments.

In the first case, as said above, we obtain the $A15$ structure, each Kelvin polyhedron, containing three Z_{14} vertices (the 12 extremities of the diagonal segments weighted by $1/4$ because each such vertex is shared by four Voronoi cells) and one Z_{12} vertex (at the centre). The structure is simple cubic rather than b.c.c., because these decorated truncated octahedra have alternate orientation at the cube vertices and at the body centres, the cubic unit cell containing therefore six Z_{14} and two Z_{12} sites. The $A15$ structure is a packing of centred icosahedra and tetrahedra. An icosahedron is joined by eight triangular faces to other icosahedra along three-fold axes of the cubic cell. The tetrahedra are glued on the remaining 12 faces of the icosahedra. A Kelvin polyhedron contains an icosahedron, but tetrahedra, having their centres on the vertices of Kelvin polyhedra, are shared by four such cells. Note that it is this case, where the retained diagonals alternate their orientation from vertex to body centre but are aligned with the $A15$ lines, that we call the 'aligned case'; see figure 7.7.

In the second case of 'antialigned' neighbouring diagonal segments, every truncated octahedron centre sees diagonals in the same orientation. There are, of course, two enantiomorphic forms of the structure compatible with a b.c.c. lattice. But this is not a possible t.c.p. structure because some distances are too short and some atoms are badly shared by two Kelvin polyhedra. If we consider the structure as a packing of icosahedra and tetrahedra, there are two kinds of misfit: icosahedra having common faces relatively rotated by π and tetrahedra divided in four by the Voronoi cells are in positions that do not allow full tetrahedra to be rebuilt. This structure is only interesting as a virtual structure on which we can apply a $\Omega(\lambda)$ operation and then get a tetrahedral close-packing structure.

S. M. and Bergman structures are obtained by one $\lambda = 3$ decoration on the $A15$ or the b.c.c. modification of the $A15$, seen as a packing of truncated

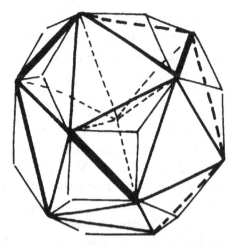

Fig. 7.7. Decomposition of the truncated octahedron into tetrahedra. Square face diagonals (thick lines) are segments of disclination lines in the $A15$.

octahedra which are tetrahedrized. The centred icosahedron is decomposed into 20 distorted tetrahedra. The remaining space in the truncated octahedra consists of 12 quarter-tetrahedra. In the $A15$ structure, quarter-tetrahedra combine into one full tetrahedron. In the b.c.c. modification, they remain as separate units, but what would have been the geometric centre of the tetrahedron is decorated by an atom of the Bergman structure as in the $\lambda = 3$ procedure tetrahedra are centred. In both Bergman and S. M. structures, every diagonal of the square faces of the truncated octahedra is the common edge to six tetrahedra, as befits a negative disclination.

Starting from the centre of the truncated octahedra, we obtain the same sequence of concentric polyhedra for both structures, since they are both decorations of the same truncated octahedra: a central vertex (Z_{12}), icosahedron of Z_{12}, dodecahedron of Z_{16}, icosahedron of Z_{12}, 'soccer ball' (truncated icosahedron) and truncated octahedra. It is important to notice that, in the two final structures, all the 24 vertices of the Kelvin polyhedra are occupied by atoms, but it is not always the same set of 12 vertices which appears in the source structure.

This shelling can be seen directly from the decoration of the truncated octahedra (Kelvin) with its enclosed topological icosahedron capped by 12 quarter-tetrahedra. First shells corresponding to the decoration of the icosahedron are topologically identical to those found in the polytope obtained by one iteration on the $\{3, 3, 5\}$ polytope. The centres of the radial tetrahedra (decorated into Laves–Friauf polyhedra) constitute the dodecahedron of Z_{16}

vertices. The disclination lines are the 30 edges of the dodecahedron, plus 20 radiating outwards, traversing the hexagonal faces of the soccer ball. Of these, eight also traverse the hexagonal faces of the truncated octahedra, and terminate on the dodecahedra belonging to nearest neighbour truncated octahedra.

The other 12 terminate on the centres of the quarter-tetrahedra (strictly, on what would be their centres if they were whole), which make up half the vertices of the original truncated octahedron. The remaining truncated octahedron vertices are the outer vertices of the radial tetrahedra; see figure 7.8.

The only difference, when Bergman and S. M. structures are compared, is in the relative orientation of nearest neighbour decorated truncated octahedra. This affects the coordination of some of the outer vertices. The soccer balls are the same in both structures: all are Z_{12} except those 12 on the $A15$ disclination segments which are, accordingly, Z_{14}. The truncated octahedron vertices divide into two classes. Half of them are on the $A15$ disclination segments. They remain Z_{14} in S. M. and become Z_{15} in Bergman. The other half are the 'centres' of the quarter-tetrahedra. They are Z_{16} in S. M. (receiving one disclination segment from every incident truncated octahedron), and Z_{15} in Bergman.

The orientation of square face diagonals in the truncated octahedron induces an orientation of the disclination dodecahedron. Nearest neighbour disclination dodecahedra are linked by disclination segments, which are eclipsed in S. M. (opposite orientation, mirror symmetry across hexagonal truncated octahedron faces) and staggered in Bergman (same orientation, inversion symmetry at the centre of hexagonal truncated octahedron faces). Figure 7.9 shows the

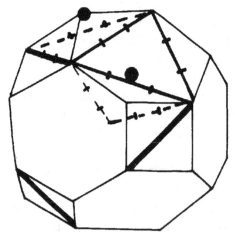

Fig. 7.8. The $\lambda = 3$ decomposition of the truncated octahedron, tetrahedrized as in figure 7.7. (Dots are centres of Friauf–Laves polyhedra.)

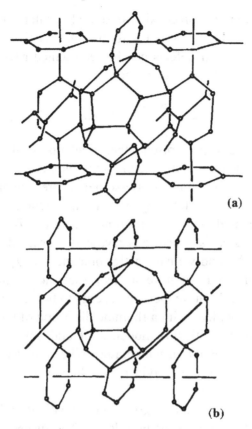

(a)

(b)

Fig. 7.9. The disclination networks (major skeletons) of Bergman (a) and S. M. (b) structures (from Rivier and Sadoc 1988).

disclination networks of these two close structures. The Bergman structure has one single network with some Z_{15} vertices. It is manifestly b.c.c. The S. M. structure contains two interlaced disclination networks, one of which is the $A15$'s, and the other an arrangement of interlinked dodecahedra on a b.c.c. lattice. The two structures are really polymorphs of one structural principle: they have the same number ($N = 162$) of vertices (atoms) per cubic cell (81 per truncated octahedron), which form two identical clusters of 81 atoms. It is only the relative orientation of the two clusters which differs. The average coordination $z = 1082/81 = 13.358$ is also the same in both structures. They have the same elastic energy density to first order, measured as the number of disclination segments per unit cell, $2E = 2n_{14} + 3n_{15} + 4n_{16} = 110$ (E, N and z are related by the identity $2E = (z - 12)N$ and are not independent). Here, n_{14}, n_{15}, n_{16} and n_{12} denote the number of 14, 15, 16 and 12-coordinated atoms.

Clearly, the two structures can coexist, separated by interfaces, since they are made of the same structural units, but with different relative orientations. The two structures and structural defects can be represented as an effective Ising hamiltonian on a body centred cubic lattice,

$$E = -J_1 \sum \langle ij \rangle S_i S_j - J_2 \sum \langle \langle ik \rangle \rangle S_i S_k \qquad (7.2.3)$$

where the Ising 'spin' $S_i = 1$ represents the orientation of the disclinations in the ith truncated octahedron, and the summation is made over nearest neighbours $\langle ij \rangle$ (in different cubic sublattices) and next nearest neighbours $\langle \langle ik \rangle \rangle$ (in the same sublattice, separated by a square truncated octahedron face). $J_2 > 0$ orders the sublattices and stabilizes uniform structures. J_1 couples the sublattices. If $J_1 > 0$, Bergman's structure is favoured over S. M. (as seems to be the case experimentally) and we have a 'ferromagnet'. If $J_1 < 0$, S. M. is the ground state structure and we have a (two sublattices) 'antiferromagnet'. Because Bergman and S. M. have the same energy to first order, $J_1 = 0$ and the two sublattices are uncoupled. With a rhombic distortion of the square faces in Bergman, J_1 is positive. With this model, structural defects can be studied quantitatively and as a function of temperature. The simplest structural defect consists of one wrong 'spin', a Kelvin polyhedron of the wrong orientation amid a uniform phase.

Here are the changes in atomic positions, coordinations and in the disclination network, associated with a structural defect. The successive atomic shells are the same in Bergman and S. M. structures. The vertices of the truncated octahedron are shared between four b.c.c. unit cells, but they are in the same positions in both structures and fit across a defect (possible deformation in Bergman of the square faces into rhombi notwithstanding). Vertices of the soccer ball lie on the faces of the truncated octahedron and are shared by two unit cells. Those on the hexagonal faces (nearest neighbour interfaces) are at identical positions in both structures, and fit perfectly. The only nearest neighbour adjustment is in the relative orientation of the disclination dodecahedra. Only vertices of the soccer ball lying on the square faces of the truncated octahedron, interfaces between next nearest neighbours, are mismatched. To patch the structure up, a T1 transformation (exchange of neighbours) (Weaire and Rivier 1984) is required on one of the soccer balls (see figure 7.10) on either side of the interface. This can be done locally, without affecting other interfaces.

The main changes are described in figure 7.11: note that two vertices are now 13-coordinated (Z'_{13}, which is non-canonical in defect-free Frank–Kasper phases). Exotic coordination, together with positive disclinations, is sympto-

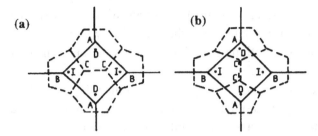

Fig. 7.10. (a) Vertices of the outer concentric shells viewed from the square face of the truncated octahedron (TO). TO: A, B, full line; soccer ball: C, dotted line; outer icosahedron: I; dodecahedron: D. (b) A T1 transformation in the soccer ball ($C \rightarrow C'$) is required to connect the TO unit cell of the wrong polymorph to the rest of the structure (structural defect). A T1 transformation is the conventional notation for the flip of an edge in a two-dimensional tiling having a coordination of 3.

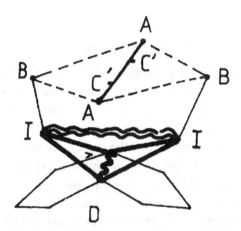

Fig. 7.11. The pelota of disclinations accompanying the structural defect of figure 7.9. Full lines: negative disclinations. Wavy lines: positive disclinations. Thin lines indicate the (negative) disclination network unaffected by the structural defect. Labels A and B, C, I and D, are the nearby vertices of the TO, soccer ball, outer icosahedron and dodecahedron, respectively. Only the two soccer ball vertices C need be moved to C'. This decreases the coordination of D and A by one and increases by one that of B and I. Note that the two I vertices are now 13-coordinated.

matic of disorder. These coordination changes must be balanced by disclination lines which should be as short as possible in order to minimize the energy. A 'pelota' (wool-ball) of disclination is formed. The additional energy is $J_2 = 6$ disclination segments per square interface, which is comparable with that of the ordered structure (54 disclination segments per Kelvin polyhedron, i.e. 18

per square face for either structure). Polycrystallinity (coexistence of the two structures) and structural defects are therefore likely to occur.

The pelota of disclinations patching up the structure can be introduced on either side of every square interface (defect), at the same cost in energy. One finds again 'two-level systems' associated with disorder.

While negative disclinations play a double role of decurving and disordering, positive disclinations are associated only with disorder.

It is interesting to note here the work by Winters and Hammack (1993) who get an amorphous alloy by applying pressure on an Al_5Li_3Cu body centred cubic alloy, with a structure close to the Bergman structure (except for empty sites at the centre of icosahedra, so with 160 atoms in the cubic cell). For 23.2 GPa pressure, there is a transition to an amorphous state which is irreversible. Such a structure can probably be described in terms of a disordered disclination network.

7.2.6 The γ phase

In 1926, as part of his thesis work, Hume-Rothery (1926) published a study on the stability of metallic alloys. It contained interesting ideas on the role of electron concentration in the stabilization of the alloy structures. In short, he remarked that, in many cases, different alloys, which have the same value of the number of electrons per atoms, have the same structure. He found a sort of universal scheme, which goes, from low to high electron concentrations, from the α phase, to the β phase and then to the γ phase. Some years later, this was given a more sound theoretical basis by Jones (1960), who showed that it corresponds to a good match between Brillouin zones and the Fermi surfaces, with the opening of pseudo-gaps in the density of states at the Fermi energy and, as a consequence, a structural stabilization.

This γ phase, which is found for example in brass, is an interesting structure in the present context, as seen below. Its prototype is Cu_5Zn_8. The Hume-Rothery ratio is $n_e/n_a = 21/13$ (note that it is a good approximation to the golden mean). Another example is the Cu_9Al_4 alloy which leads to the same Hume-Rothery ratio (counting Al as trivalent).

The γ phase structure is often presented as 'complex'. We give here a new point of view based on its relation with the {3, 3, 5} polytope. The crystallographic cell is body centred cubic and is filled with two clusters of 26 atoms. The clusters are built up successively of an inner tetrahedron, an outer tetrahedron, an octahedron and a distorted cubo-octahedron whose square faces are replaced by approximately golden rectangles.

Now, a remarkable fact is that this sequence of concentric shells is also found in a {3, 3, 5} polytope around a basic tetrahedral cell, but with exact golden rectangular faces on the distorted cubo-octahedron (see table A5.2). Therefore, each cluster of the γ phase can be considered as a part of a {3, 3, 5} polytope, slightly distorted by projection, and such that the projection pole is located at a tetrahedron centre. In this case the local symmetry is tetrahedral as opposed to the local icosahedral symmetry of the Bergman–Pauling phase. Since it arises from mapping a piece of the {3, 3, 5} polytope, this means that the cluster is a tetrahedral packing. Figure 7.12 presents these shells.

Considering the structure obtained by reproducing a cluster at each node of the body centred cubic lattice, we have to analyse the connection between close clusters. Here again we can define a tetrahedrization of the space separating clusters, but there are large departures from the ideal regular tetrahedron. Then the structure can be analysed in terms of a tetrahedral packing, in which disclination lines are on edges which are not common to five tetrahedra.

The atomic cluster can be defined in the Delaunay–Dirichlet region of the b.c.c. lattice which is the interior of a Kelvin polyhedron (a truncated octahedron with hexagonal and square faces).

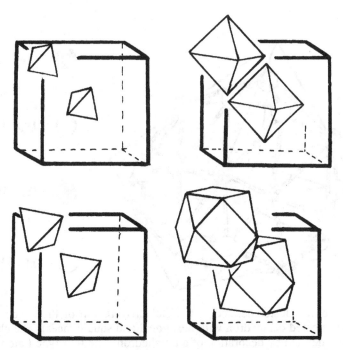

Fig. 7.12. Shells of the cluster of the γ phase centred around the cell centre. Shells appear in different cells for a best visualization.

The cluster can be schematically defined by decoration of a tetrahedron (figure 7.13).

Three type IV atoms are placed in each triangular face of these tetrahedra, forming a triangle with the same orientation as the face. Consequently vertices of the tetrahedron are surrounded by three type IV atoms in close contact which form a triangle (smaller than that which is in a face) and edges of the tetrahedron are parallel to the sides of the rectangle formed by four type IV atoms. This decoration of the tetrahedron defines the distorted cubo-octahedron formed by type IV atoms.

Type III atoms are on the edges of the tetrahedron. These form the type III octahedral shell; schematically they are at the centre of the rectangular face of the distorted cubo-octahedron (they are not exactly in the plane of the rectangle). If edges are drawn between atoms in close contact, these type III atoms are connected to type IV atoms by drawing (only approximately since they are not in the rectangle plane) the diagonal of the rectangle of the distorted cubo-octahedron.

Type II atoms complete the triangulation of the polyhedron which is the surface of the cluster. They are at the centre of the hexagon formed by three

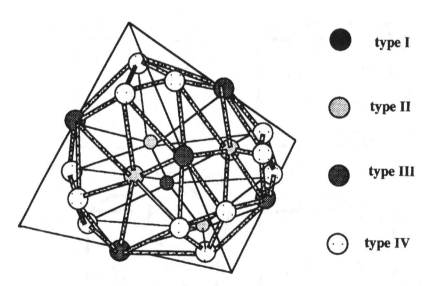

● **type I**

◉ **type II**

● **type III**

⊙ **type IV**

Fig. 7.13. The atomic cluster defined in the Delaunay region of the b.c.c.: a Kelvin polyhedron (a truncated octahedron with hexagonal and square faces). The cluster can be schematically defined by decoration of a tetrahedron. The four type I atoms form a small tetrahedron inside the cluster, but for clarity they have been omitted from the figure.

type IV atoms and three type III atoms (this is also the centre of triangular faces of the tetrahedron).

In fact the cluster described like this has all its atoms on its external surface except the four type I atoms which are inside the cluster.

This cluster can be positioned relative to the Kelvin polyhedron. Small triangles formed by type IV atoms are in four hexagonal faces of the Kelvin polyhedron. Type II atoms are inside the Delaunay cell, in front of the four remaining hexagonal faces. Type III atoms are inside the Delaunay cell in front of the square faces. When Delaunay cells are packed together in order to rebuild the structure, a hexagonal face which contains three type IV atoms is placed on a hexagonal face which is over a type II atom.

Coordination polyhedra for the four types of atom are defined by neighbours belonging to the same cluster as the considered atoms and by neighbours belonging to other clusters (see figure 7.14).

We first consider neighbours inside the cluster containing the centre of coordination polyhedra.

- Type I (small tetrahedron) atoms. The neighbours are three type I, three type II, three type III and three type IV all in the same cluster. The coordination polyhedron is an icosahedron.
- Type II (large tetrahedron) atoms. The neighbours are three type I, three type III and three type IV. So there are nine neighbours in the cluster forming an incomplete icosahedron.
- Type III (octahedron) atoms. The neighbours are one type I, one type II, two type III and two type IV. So there are six neighbours forming a pentagonal pyramid extracted from an icosahedron.

Except for the type I coordination polyhedron which is an entire icosahedron enclosed in the cluster, other coordination polyhedra have to be completed by atoms belonging to other clusters.

When two Delaunay cells enclosing two clusters are joined by a hexagonal face, the coordination polyhedron of a type II atom, in the first cluster is completed by a small triangle (belonging to the second cluster). The type II coordination polyhedron is completed into an icosahedron.

A type III atom inside the Delaunay cell is in front of a square face of the Kelvin polyhedron which borders the cell. Consider four hexagonal faces belonging to several Kelvin polyhedra which have a common edge with the square face. There are four type IV atoms in these faces which are neighbours of the type III atoms. These four atoms would complete the coordination polyhedra to form an icosahedron. But there is still another neighbouring atom of type III which is symmetric with the one considered as the centre of the

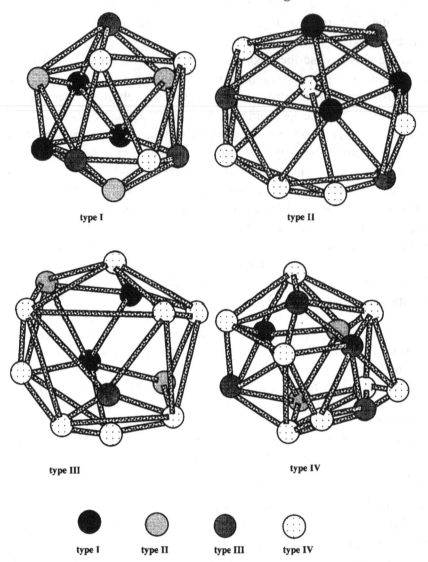

type I

type II

type III

type IV

type I   type II   type III   type IV

Fig. 7.14. Coordination polyhedra for the four types of atoms: I, II, III and IV. Coordination numbers are 12, 12, 13 and 15.

coordination shell, relative to a square face of the Kelvin polyhedron. The coordination polyhedron can be obtained from an icosahedron by adding a thirteenth atom. For that two triangular faces are changed into a square, and a square pyramid is capped on this square. This coordination polyhedron has 13 vertices.

A type IV atom has six neighbours belonging to its cluster. Then there are five other neighbours (four type IV and one type II) coming from seven neighbouring clusters. This completes the coordination polyhedron to an icosahedron less one vertex, so there is a free pentagonal face. Then this face is capped by one triangulated object which has four new vertices (two type IV and two type III). So the coordination number is 15.

The mean coordination number can be deduced from these four types of coordination. It is $\bar{z} = 12 + 21/13$ or $\bar{z} = 13.615$. Here again the coordination number is close to 13.4 corresponding to the statistical honeycomb of Coxeter, but in the present case, as we describe below, there are positive disclinations which have the effect of increasing the coordination number. This point is presented in appendix A6 and in §4.3.3.

It is possible to define disclinations from these coordination polyhedra that are completely triangulated. A disclination segment through an atom merges onto an atom of its coordination shell which does not have five neighbours on the surface of this coordination polyhedron. If this number of neighbours is six, that defines a negative disclination (which contributes to the decurving) and if there are four neighbours, that defines a positive disclination (which recurves the structure). The structure is not exactly a canonical Frank and Kasper structure: there is a more intricate network of disclination (see 'pelote; in appendix A6). But it obeys the conservation rules defined in chapter 4. It is possible to define the disclination network and its topology in a simplified way.

Consider two interlaced cubic networks in which vertices of one network are the centre of cubes of the other network (figure 7.15). Each square face of a cube is pierced by an edge of the other network. In such a square face there are four type IV atoms which form a small square oriented like a face of the network. These four points are joined by four edges belonging to the disclination network. These disclinations are positive. This square arrangement is an approximation: the four type IV atoms are the four atoms which are said, in the above description of the type III coordination, to complete this coordination shell to an icosahedron, but are capped by a pyramid leading to a coordination 13. Actually they form a screw 4-gon. But the topology is easier to describe supposing that it is a square. Symmetrical to this square there is a pair of two type III atoms, also close neighbours. The segment joining them is a positive disclination. There are from each of these type III atoms two negative disclinations going towards two opposite vertices of the type IV square. Two close type IV squares have non-intersecting orthogonal edges whose four ends form a tetrahedron. Edges of this tetrahedron (except square edges) are negative disclinations. This completes the disclination network.

Table 7.1. *Coordination number in metallic structures. Notice that the two numbers $\bar{z}$ and $\bar{n}_t$ are related by $\bar{z}(6 - \bar{n}_t) = 12$*

	Number of atoms in the cell					Mean coordination	Tetrahedra per edge
Structures	n_{12}	n_{13}	n_{14}	n_{15}	n_{16}	$\bar{z}$	$\bar{n}_t$
β-W	2	0	6	0	0	13.50	5.1111
Laves phase	8	0	0	0	4	13.333	5.1000
β-U	10	0	16	4	0	13.466	5.1122
α-Mn	24	0	14	0	10	13.517	5.1089
γ-phase	16	16	0	24	0	13.615	5.1186
Bergman phase	98	0	12	12	40	13.358	5.1016
S. M.	98	0	18	0	46	13.358	5.1016
b.c.c.	0	0	2	0	0	14	5.1414

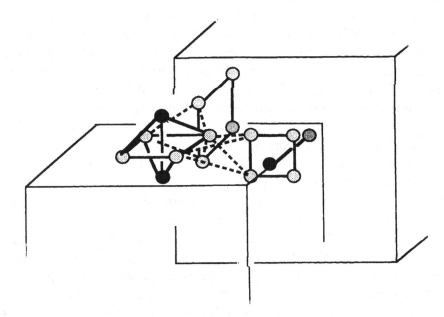

Fig. 7.15. The disclination network.

7.2.7 To recap: coordination numbers in these phases

Table 7.1 displays numerical values for the examples presented above.

All these structures are supposed to be completely tetrahedrized with atoms on tetrahedron vertices. These tetrahedra are not regular, but there are always 5.1 sharing an edge. The coordination number $\bar{z}$ is close to 13.5. Largest values correspond to phases containing positive disclinations, and consequently there are more disordered than classical Frank and Kasper phases.

7.3 Melting of model structures

Using classical molecular dynamics simulations combined with Voronoi tessellation it is possible to study the geometrical modifications as a function of temperature in models of Frank–Kasper phases (Jund *et al.* 1997). This is done by combining molecular dynamics simulations and the Delaunay–Voronoi tessellation to follow dynamically the evolution of the geometrical characteristics of the Voronoi cells as a function of temperature, in order to shed new light on the melting of Frank–Kasper phases from a *geometrical* point of view. Then it is possible to follow the behaviour of the major skeleton formed by the Frank–Kasper lines, which are also disclination lines.

We consider here two model Frank–Kasper phases: the $A15$ structure (β-tungsten) and the cubic Friauf–Laves structure (MgCu$_2$) presented in the previous section. These structures are not only the simplest but also correspond to the lowest and the highest coordination number in Frank–Kasper structures.

Their high icosahedral order is reflected best in the geometrical characteristics of the Voronoi cells. The evolution of the Frank–Kasper lines with temperature is the most efficient way to characterize how the icosahedral order is affected by temperature, or in other words how these Frank–Kasper phases melt.

7.3.1 Method and model systems

Molecular dynamics simulations are done for microcanonical systems of soft spheres interacting via the inverse sixth power potential defined by Laird and Schober (1991):

$$U(r) = \epsilon \left(\frac{\sigma}{r}\right)^6 + Ar^4 + B$$

in a rigid cubic box of edge length L, with periodic boundary conditions at constant density.

This potential has been chosen by analogy to previous work on glasses. It is an efficient potential, when applied to constant volume simulation, in order to get glasses after quenching the liquid. Clearly, it is not at all suited to simulating true metallic Frank–Kasper phases, but is a good tool for a geometrical study including all related structures like clathrates.

Starting from initial samples at 0 K, the following configurations were obtained by annealing this sample with an annealing rate of 10^{12} K/s which was achieved by removing the corresponding amount of energy from the total energy of the system at each iteration.

At several temperatures during this annealing process, configurations (positions and velocities) were saved. Each configuration was then used to start a constant-energy molecular-dynamics calculation during which the temperature evolves towards a saturation regime. Then, geometrical quantities are calculated during additional steps of the molecular dynamics simulation.

In fact during these additional steps, the molecular dynamics scheme and the so-called 'Voronoi–Delaunay tessellation' are combined. A Voronoi cell, as the ensemble of points closer to a given atom than to any other, is obtained by an efficient algorithm which has been recently introduced to study large random sphere packings (Jund _et al._ 1997). The first step of this algorithm is to determine the Delaunay tetrahedral simplicial cells for a given configuration. These cells are among all the tetrahedra formed with four atomic centres, such that no other atomic centre lies inside their circumscribed spheres. Once the simplicial tetrahedra are obtained, the Voronoi cell of a given atom is determined knowing that its vertices are the centres of the circumspheres of all the simplicial tetrahedra sharing this atom. During the simulation it is possible to record the geometrical characteristics of these cells, among them the mean number of faces $\langle F \rangle$, also called coordination number z, as well as the fractions f_e of cell faces having a given number e of edges ($e \geqslant 3$). Then in this approach, a non-pentagonal face of the Voronoi–Delaunay decomposition corresponds to a 'defect' and the corresponding face then encircles a segment of a 'disclination line'.

Frank–Kasper lines are disclination lines appearing on bonds shared by six tetrahedra instead of five for the perfect icosahedral order. By definition, Frank–Kasper structures contain only this kind of disclination, but other structures can contain other types of disclination, for example when a tetrahedron edge is common to four or three tetrahedra. Upon heating, such defects appear.

Frank–Kasper lines have to respect conservation rules. In fact these lines behave as if they have a line tension proportional to their angular deficit $\delta_e = 2\pi - e\alpha_t$, where e is the number of tetrahedra sharing an edge and $\alpha_t = \cos^{-1}(1/3)$ is the tetrahedron dihedral angle.

In the $A15$ structure all lines, which never intersect, are straight lines. In Friauf–Laves phases four segments ending on a node have a highly symmetric configuration. When positive and negative disclinations are mixed they have to respect such rules: some examples are described below for disclination lines appearing upon heating.

There is also a balance for the space curvature. In order to consider this problem it is helpful to suppose that the structure is not lying in a flat space but in a corrugated one, flat only on average. Curvature concentrations are then located on tetrahedron edges and they are positive for edges shared by three, four or five tetrahedra and negative for edges shared by six or more. In three dimensions there is no exact relation between the values of f_e resulting from the global vanishing curvature. Nevertheless there exists an approximate relation $\sum_e \delta_e f_e \simeq 0$ (Sadoc and Mosseri 1984). This gives, if we consider that there are only pentagonal and hexagonal faces for Voronoi cells, $f_5 = 0.8958$ and $f_6 = 0.1041$. In the Frank–Kasper structures studied here f_5 and f_6 are close to these values.

7.3.2 The heating process

We follow the evolution of the local structure during the heating process. Figure 7.16 shows the variation of f_e as a function of temperature (once the system has reached equilibrium) for the β-W structure and the MgCu$_2$ structure. As expected for a solid–liquid transition, at the transition point the physical parameters are discontinuous, which is typical of a first order transition. In particular a sharp drop in f_5 can be observed at $T_m \approx 13.5$ K (β-W) and $T_m \approx 8.0$ K (MgCu$_2$) indicating the passage from solid to liquid behaviour. It is worth mentioning that these curves are basically unchanged when, for example, a faster annealing rate is used (10^{13} K/s). This indicates that these curves are indeed equilibrium curves. The breakdown of f_5 towards the liquid value of about 0.4 (and the corresponding increase of all the other f_e) happens

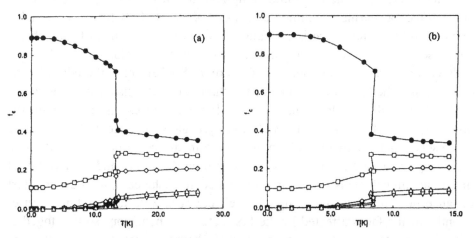

Fig. 7.16. (a) Temperature evolution of f_e for β-W: ∇: $e = 3$; $\diamond$: $e = 4$; $\bullet$: $e = 5$; $\square$: $e = 6$; $\triangle$: $e = 7$. (b) Same as (a) for MgCu$_2$.

for both structures when $f_5 \approx 0.7$, which indicates that the breaking mechanisms of the icosahedral order are probably similar.

Note that in figure 7.16b the temperature (8.1 K) of the first liquid configuration is slightly lower than that (8.5 K) of the last solid configuration. This is a consequence of our simulations being done in a microcanonical ensemble and not in a canonical one.

Figure 7.17 gives a better idea of what really happens to the structure as the temperature is increased. The two-dimensional projection of the network of disclination lines connecting 'hexagonal' faces (these are the only lines present at low temperature) is shown at three representative temperatures for the two structures. In figures 7.17a and 7.17d we show a snapshot of the network at the temperature for which the first defect appears. These defects look very similar even though the two structures are quite different and we will discuss this point in the following section. In figures 7.17b and 7.17e we show the disclination lines just before the melting transition, while in figures 7.17c and 7.17f the lines are shown just after the transition. It is striking that in the solid phase, even though the disorder is quite important, the underlying periodic network of disclination lines is still present and visible while in the liquid phase it has completely disappeared.

The first defect observed in both cases (figures 7.17a and 7.17d) is of the same kind and can be discussed as follows. Let us consider two simplicial tetrahedra ($A_1A_2A_3A_4$ and $A_1A_2A_3A_5$; see figure 7.18) sharing a common triangular face ($A_1A_2A_3$). If the distance (A_4A_5) between the non-common atoms becomes sufficiently small (so that one enters the circumspherical sphere of the other tetrahedron) they become nearest neighbours and a new Voronoi cell face is created. The two initial simplicial tetrahedra are replaced by three new tetrahedra sharing a common edge ($A_1A_2A_4A_5$, $A_2A_3A_4A_5$ and $A_3A_1A_4A_5$). Subsequently the new face is a triangle formed by the centres of the spheres circumscribed to these tetrahedra. The change in the topology of the Voronoi cells is summarized at the bottom of figure 7.18 where only the edges of the cells are sketched. As a consequence the three faces associated with the bonds A_1A_2, A_2A_3 and A_3A_1 lose one edge, while the six faces associated with the bonds A_4A_1, A_4A_2, A_4A_3, A_5A_1, A_5A_2 and A_5A_3 gain one edge. In the $A15$ case we observe that the two initial simplicial tetrahedra corresponding to the first appearing defect are such that A_4 and A_5 are on the centre and on a vertex of the cubic unit cell. This is not surprising since the distance A_4A_5 is the smallest distance not present in the simplicial cell edges at zero temperature. Consequently all the faces affected by the first defect are pentagons so that together with the new triangle, three squares and six hexagons are created while nine pentagons disappear. In the cubic Friauf–Laves case we observe that the two

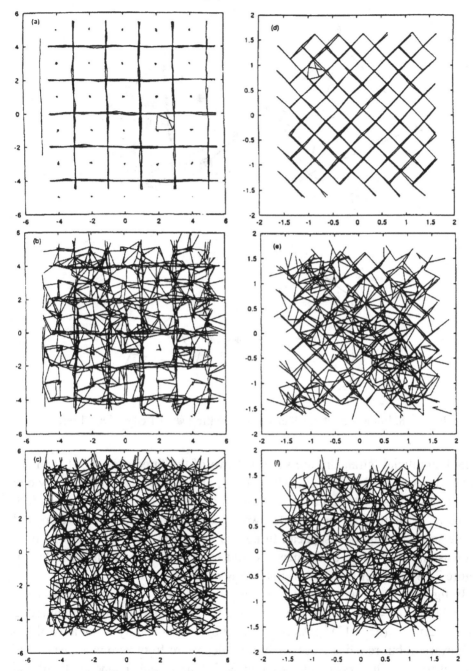

Fig. 7.17. Disclination lines for β-W: (a) $T = 3.5$ K, (b) $T = 13.4$ K, (c) $T = 13.7$ K. Disclination lines for MgCu$_2$: (d) $T = 1.9$ K, (e) $T = 8.5$ K, (f) $T = 8.1$ K.

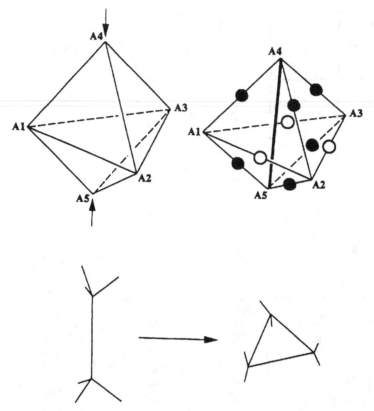

Fig. 7.18. Initial defect mechanism (schematic): ●: +1 edge; ○: −1 edge.

initial simplicial tetrahedra are such that two of the three sites A_2, A_3 and A_4, for example A_2 and A_3, are Z_{16} while A_1, A_4 and A_5 are Z_{12}. Consequently one face (bisecting A_2A_3) is a hexagon while all the others are pentagons; therefore globally, in addition to the triangle, two squares and five hexagons are created while seven pentagons disappear. In figures 7.17a and 7.17d one sees the two-dimensional projection of the new six-fold bonds (indicated with black dots in figure 7.18) together with the original array of disclination lines.

In figure 7.19 we have monitored the quantity $c = \sum_e f_e \delta_e$. Note that, using the relation $\langle e \rangle = (6z - 12)/z$, it is related to the mean coordination number z by $c = 2\pi - \alpha_t(6z - 12)/z$. As indicated above, this quantity is not a topological invariant but remains close to zero. Its departure from zero is a measure of the disorder: it has been discussed in §4.3 that a high connectivity of the disclination network increases c, while asymmetry of the Voronoi cells, or positive disclinations, decrease c. The increase of the thermal disorder by heating as well as the passage from solid to liquid contribute to lowering the value of c.

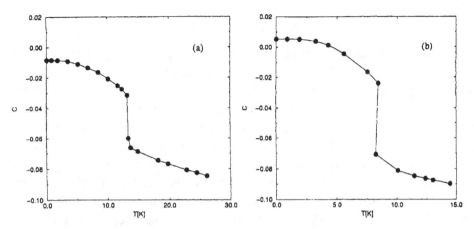

Fig. 7.19. The quantity $c = \sum_e f_e \delta_e$ decreases with temperature, but remains close to zero. This evolution follows the decrease of the icosahedral order, with a jump at the melting temperature. The behaviour is similar for the two structures (Frank–Kasper (a) and Friauf–Laves (b)), except for positive values, at low temperatures for the Friauf–Laves structure related to the high connectivity of the disclination network.

It is interesting to note that because all the faces were initially only pentagons and hexagons, all the defects can be detected by counting the number of triangular faces present in the system at least up to a temperature such that these defects do not start to interact. Since a new bond is created per defect the increase of z, $\Delta z = z - z_0$, should be equal to the number of triangular faces per atom, which is $z f_3$, that is

$$\frac{z - z_0}{z} = f_3$$

This relation is fulfilled at low temperature as long as the defects do not interact. Note that the creation (annihilation) process of these defects is a quite natural extension of the so-called T1 transformation previously introduced in two dimensions (here there is no analogue of the T2 transformation as there is a natural conservation of the number of cells) (Weaire and Rivier 1984). This sheds new light on the importance of small triangular faces as a measure of disorder in three-dimensional structures.

Finally, close to the melting temperature, not only do these defects interact but certainly other kinds of defects start to exist. It is interesting to consider what happens upon cooling. If the melt is cooled, as usual with this potential a glassy structure is obtained, but if the structure just below the melting temperature is cooled the perfect structure is recovered. Few traces of the initial disclination network are sufficient to rebuild the structure.

7.4 Tetracoordinated structures

In covalent structures like a-Si or a-Ge the short range order is perfectly defined
by tetrahedral connectivity: one atom at the centre of a tetrahedron is linked to
its four neighbours by bonds directed towards the vertices of the tetrahedron.
As this local order may lead to several different structures in flat or curved
space it is necessary to add some details about the order at a scale slightly
larger than the first neighbour distances. The important topological ingredient
for improving the description of the local order is the local ring configuration
(the number of edges in the ring and its parity, the twist of the ring and the
existence of cages). In dense and tetracoordinated structures the similarity of
the local symmetry (tetrahedral interstices and bonds) leads to a relation
between the two types of structure and this explains why we can obtain models
for covalent structures starting from the $\{3, 3, 5\}$ polytope and others derived
from it, as described in the first chapter.

7.4.1 Clathrate structures

Ice associated with other molecules, such as a rare gas, is the best known
example of clathrate. In this case the tetracoordination holds to the electronic
structure of the oxygen atom of one water molecule, leading to the formation
of two covalent bonds with its protons and two hydrogen bonds with those of
two other water molecules.

There are two types of structure which describe most of the clathrate
hydrates (Davidson 1973) and which appear to be dual of the $A15$ Frank and
Kasper structure (type I clathrate) and of the Laves phase structure (type II
clathrate). In these structures the water molecules build cages (one can consider
figures 7.24 and 7.25), with oxygen atoms at the vertices and covalent and
hydrogen bonds along the edges, trapping the associated molecules. These
structures can be analysed as $\{5, 3, 3\}$ polytopes, the dual of the $\{3, 3, 5\}$,
which have been flattened by disclinations, the disclination networks being
similar to that of the β-tungsten structure in the case of type I clathrates or to
that of the Laves phase structure in the case of type II clathrates.

There is another example of very interesting clathrates which complement
the many types of structure formed by Si. No closed cages are observed in the
crystalline form; they most likely do not exist either in the amorphous material,
but the compound $Si_{1-x}Na_x$ exhibits clathrate structures with cages built by the
Si atoms and trapping the Na atoms when $0.02 < x < 0.2$, type I for large x and
type II for small x (Cros *et al.* 1965).

As well as the $A15$ and Laves phase structures there exist other structures which can be described following the same approach, for instance β-U or Frank and Kasper alloys; there exist also other clathrate structures with different connectivities of their disclination lines, for instance three in bromide hydrate instead of two in type I or four in type II.

7.5 Liquid crystal structures

7.5.1 Periodic structures of disclinations in 'blue phases'

These are assemblies of elongated organic molecules, without any long range translational order, but with a long range orientational order corresponding to a rotation of the mean molecular axis for any displacement orthogonal to it. As shown in chapter 2, such an orientational order cannot propagate without frustration in our Euclidean space but, nevertheless, crystalline organizations called 'blue phases' are observed. To describe their ideal structure, we make use of a particular set of lines in S^3: Clifford's parallels in S^3. In order to relate the ideal structure in S^3 and real structures in R^3, defects must be introduced to change the space curvature (Pansu *et al.* 1987). Blue phases are, from this point of view, similar to Frank and Kasper phases, but with a continuous frustration.

Another approach consists of describing directly these phases in R^3 with defects (Meiboom *et al.* 1983). Finite size domains, in which the double twist is not perfect but relatively well approximated, are arranged on a cubic lattice. In these 'double twist tubes', with cylindrical symmetry, molecules have an azimuthal component which increases with the distance to the axis of the cylinder (figure 7.20). Then, blue phases are described by a periodic arrangement of double twist tubes with cubic symmetry (for example the blue phase II structure in figure 7.20) following usual space groups observed for blue phases.

There are disclination lines which are associated with these arrangements of tubes and form a network of defects. In order to see that, consider some [1 1 1] axes, and the director field surrounding these axes. As seen in figure 7.21b, following the field rotation along a loop surrounding the axis, there is a π disclination. This disclination belongs to a disclination network associated with the periodic double twist tubes arrangement. There are other [1 1 1] directions which are not defects (figure 7.21a); they can be seen as the axis of another family of double twist tubes, so the description in terms of tubes is not unique. For instance the blue phase II can be described by the arrangement of figure 7.20, but also by an arrangement of tubes following the axes of figure 7.21b. The blue phase I can be described with tubes along cubic axes, or with a tube network related to that of figure 7.23c.

This is a new view, in which the two blue phases I and II appear as a network

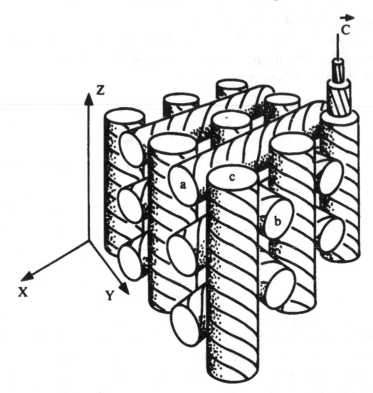

Fig. 7.20. A model of blue phase structure: periodic arrangement of double twist tubes modelling the blue phase II structure.

of double twist tubes along the skeleton used to define the P and D infinite periodic minimal surfaces. These surfaces have the same symmetry group as blue phases (we exclude mirror reflections forbidden in systems containing chiral molecules). The infinite periodic minimal surfaces allow us to see the director field between the double twist cylinders. A local minimum of the energy is obtained if the director field is following asymptotic lines of the surface (Dubois-Violette and Pansu 1992). Flat points on the surface, which are a singularity for this field, are along the [1 1 1] direction followed by disclinations. One can see in figure 7.22, a director field engraved on the P surface and imagine what occurs on close surfaces (Dubois-Violette and Pansu 1988, Dubois-Violette *et al.* 1990).

7.5.2 Crystals of films

Some liquid crystals are described as periodic organizations of two fluid media separated by interfaces (Charvolin and Tardieu 1978); they most often occur as

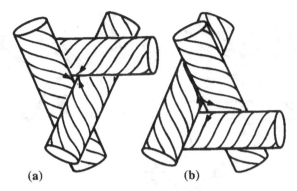

(a) (b)

Fig. 7.21. A model of the blue phase structure: disclinations appear when one tries to fill in between the cylinders. (a) In this case there is no defect, the vector field is rotating like the tubes. (b) In this case there is a defect: the two rotations within the cylinder and between the cylinders are opposite.

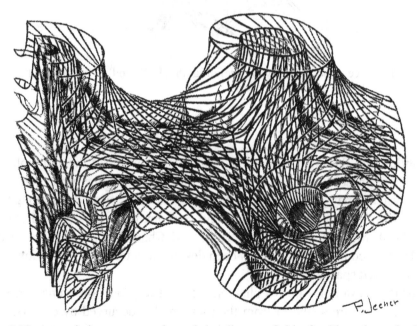

Fig. 7.22. An artistic representation of the director field of a blue phase, by Patrice Jeener.

stackings along one dimension of fluid layers of molecules, the so-called smectic phases. In the case of the lyotropic liquid crystals, built by amphiphilic molecules such as soaps, detergents and lipids, they are called lamellar phases and can be described as stackings of alternating layers of water and

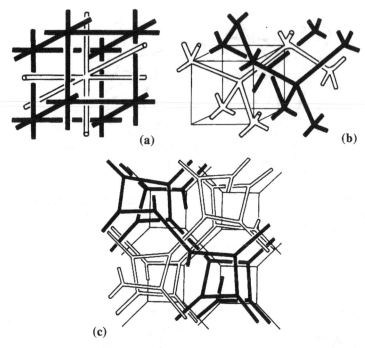

Fig. 7.23. The three types of network defining the P, D and G surfaces: they are equidistant from black and white networks.

amphiphiles with flat interfaces defined by the polar heads of the amphiphiles. Besides these phases with periodicity along one dimension, the phase diagrams of these systems may present other ordered phases with periodicities along two or three dimensions, curved interfaces and complex topologies (Ekwall 1975, Fontell 1990, Mariani *et al.* 1988, Sadoc and Charvolin 1986, Clerc *et al.* 1991). In the second chapter this has been described in terms of frustration, and solved in curved spaces S^3.

As for other examples in this chapter, we want to show how some periodic geometries in R^3 are analysed from the ideal model, decurved by defects. We present the case of cubic structures, observed with two kinds of topologies: 'bicontinuous' and 'micellar'.

A few phase diagrams of amphiphilic molecules in the presence of water exhibit cubic structures in two distinct concentration domains (Charvolin and Sadoc 1988b). Some of them are observed in the vicinity of the lamellar phase, they are now well characterized, their space groups being Ia3d or Pn3m in most cases, Im3m in some others. They can be described as being built by two interwoven labyrinths of one medium separated by a film without self-intersection of the other, as shown in figure 7.27, hence the characterization of their topology as 'bicontinuous'. The other cubic phases are observed in the vicinity of the micellar phase, their space groups are Pm3n, Fd3m or P4₃32 (Balmbra *et al.* 1969, Charvolin and Sadoc 1994). It is expected that their structures are built with finite micelles of one medium separated by a self-intersecting film of the other, as shown in figures 7.24 and 7.25. These structures are very close to the clathrate structures.

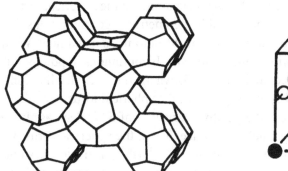
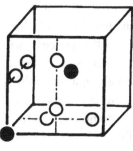

Fig. 7.24. Periodic aggregation of slightly distorted 12-hedra and 16-hedra. On the right: position of these polyhedra in the cubic cell; they are at positions occupied by atoms in the β-tungsten structure.

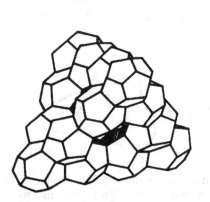
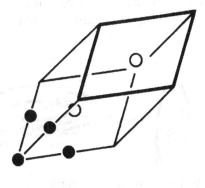

Fig. 7.25. Periodic aggregation of slightly distorted 12-hedra and 14-hedra. They are at positions occupied by atoms in the cubic Laves phase.

7.5.3 Generation of the bicontinuous topology

From the properties of S^3 and its family of parallel tori, it appears that the set of the spherical torus surrounded by two parallel tori at equal distances on either side of it can be considered as a representation without frustration of the periodic system of fluid films which is frustrated in Euclidean space. The periodicity in this space, which is the repetition of the films when moving along the geodesics of the space normal to the films, is reproduced in curved space by the displacements along great circles of S^3 normal to the tori. Moreover, as already said above, the two subspaces separated by the spherical torus are identical, as are the two spaces separated by the film in the cell of the structure and, finally, the frustration is obviously relaxed as the two tori parallel to the spherical torus have a smaller area than it has. The possible geometrical configurations in our Euclidean space are to be found by mapping S^3 onto it, as usual now. We limit ourselves to the process which maintains the bicontinuous topology of the spherical torus, as suggested in figure 7.26, where one $-\pi$ disclination was introduced along a two-fold symmetry axis of the spherical torus.

In the course of this process the positive curvature of S^3 is decreased to zero in order to come back to the Euclidean space. Then, the spherical torus, which is a surface with zero Gaussian curvature in S^3 (this intrinsic curvature is defined in the appendix A1 §4), becomes a surface with negative Gaussian curvature in the Euclidean space, as are the surfaces separating the two labyrinths of the cubic structures. The possible symmetries of these surfaces are also imposed by those of the spherical torus and the disclinations.

7.5.4 Generation of the micellar topology

In this case we use, as a supporting surface for the film of amphiphiles in S^3, the second surface separating S^3 into two identical subspaces. This surface is a

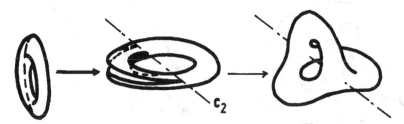

Fig. 7.26. How a torus is transformed by a $-\pi$ disclination into a genus two torus. The intrinsic curvature of this surface is thereafter negative. This figure gives mainly the topological aspect of the transformation, for a more realistic view see figure 7 in Charvolin and Sadoc (1987).

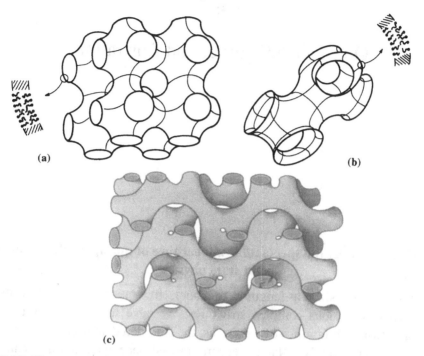

(a)

(b)

(c)

Fig. 7.27. Three examples of crystals of films: (a) based on the P-surface, (b) the D-surface and (c) the G-surface.

great sphere S^2, which can be considered as the exact equivalent in three dimensions of the equator drawn on a two-dimensional sphere and separating it into two hemispheres. Just as, on the two-dimensional sphere, there is a family of smaller circles parallel to the equator with their centres at the poles, there is, in the three-dimensional sphere S^3, a family of smaller spheres parallel to S^2 with their centres at the poles of S^3.

Thus, the great sphere S^2 with two parallel spheres at equal distances on either side of it can be considered as another representation without frustration of the periodic system of fluid films which is frustrated in Euclidean space. The periodicity in this space, which is the repetition of the films when moving along the geodesics of the space normal to the films, is reproduced in curved space by the displacements along great circles of S^3 normal to S^2. Moreover, as already said above, the two subspaces separated by the great sphere are identical, as are the two spaces separated by the film in the cell of the structure and, finally, the frustration is obviously relaxed as the two spheres parallel to the great sphere have a smaller area than it has.

8

Quasiperiodic order and frustration

8.1 Quasicrystals: the spectacular appearance of quasiperiodic order in solid state physics

As has been emphasized at length in this book, frustration most often leads to the presence of various complex structures. We have so far concentrated on essentially two such families of structures, the disordered amorphous ones, and the large cell crystals. But Nature recently proved once again that she can use all possibilities to fill space that mathematics allows. We already knew that almost all the three-dimensional space groups describe at least one real structure. But, more than ten years ago, solid state physicists received a great surprise (if not a great shock for some): the experimental result that certain metallic alloys, which were quickly called quasicrystals, adopt a long range icosahedral order at the atomic level (Shechtman *et al.* 1984), whose signature is their diffraction spectra presenting peaks – indicating long range order – displaying icosahedral order – a forbidden symmetry for standard crystallo-graphy. A new and very active field of research was born, which could take advantage of the fact that, on the one hand, and rather rapidly, thermodynami-cally stable materials of high quality have been synthesized; and, on the other hand, from the theoretical point of view, that quasiperiodic plane tilings, proposed in the mid-1970s by Roger Penrose, quickly gave an approximate image of the atomic arrangement in these alloys, through their generalization in three dimensions.

We do not aim at giving here a review of quasicrystal physics. Interested readers have a choice among many review papers and books on that subject (Janot 1992, DiVincenzo and Steinhart 1991, Hippert and Gratias 1994). We shall here focus on certain aspects of quasiperiodic order which are more linked to the main subject of this book. Geometrical frustration appears when a local order cannot propagate freely in space. The unfrustrated systems, which are free of defects, are associated with regular arrangements, periodic in space. So, a question is immediately raised: what about quasiperiodic order? Should we classify it on the side of unfrustrated structures, arguing about the positional and orientational order manifested by punctual diffraction spectra? Or, on the contrary, and it is our choice here, should we say that it represents a partial solution, and probably the most ordered possible, to the frustration problem? One should stress here that we only consider those quasiperiodic structures which present an orientational order, that is a very small class among the *a priori* possible structures, but a large class among those which have been observed up till now.

It is precisely in terms of partial solutions to frustration, and therefore as ideal models for amorphous solids, that we first became interested in Penrose tilings and proposed, after Kramer (1982), three-dimensional quasiperiodic tilings with icosahedral symmetry. Since amorphous metals are well modelled by polytetrahedral structures, we were interested in finding quasiperiodic tilings built with tetrahedral basic units (Mosseri and Sadoc 1982). These structures were later put in the standard language of quasiperiodic tilings obtained from higher dimensions (Kramer and Papadopoulos 1994). It is probably a similar line of thought, but with the aim of describing supercooled liquids, which brought Levine and Steinhart (1984) to being interested in three-dimensional Penrose-like structures. The experimental discovery of quasicrystals qualitatively changed the whole problem. Not only are we dealing with ideal structures, but it is now real materials which enter the game.

Note that, at the beginning, serious, and sometimes justified, remarks were raised about the reality of this new quasiperiodic order, arguing for instance about the metastable character of the phases, and the permanent presence of defects. But those who believed in its reality worked hard in improving the material quality – finding stable phases – and the experimental observation procedures in such a way that they finally could convince the solid state physics community.

During the past few years, the structural quality of quasicrystals, associated with the generation of new alloys (AlMn, then AlLiCu, then AlCuFe, then ...), is such that this type of order competes, as regards its perfection, with the best known crystals.

The rather quick development of the structural analysis of these phases is partly due to the very powerful geometrical method, using high-dimensional spaces, developed simultaneously in three places (Duneau and Katz 1985, Elser 1985, Kalugin *et al.* 1985). Those few who took part at the colloquium on mathematical crystallography, held at the 'Institut des Hautes Etudes Scientifiques' (at Bures sur Yvette, near Paris) in January 1985, recall the invited talk by Denis Gratias telling them about the discovery of quasicrystals and stressing the many questions raised by the analysis of diffraction spectra; after the talk, during the part devoted to questions, Michel Duneau and André Katz were able to explain most of these mysterious aspects with their 'cut and projection' method. This method is briefly recalled in appendix A7; it allows us in particular to stress the very peculiar relation between quasiperiodic structures and large cell crystals (like those studied in chapter 7).

In a first step, we describe briefly how the quasiperiodic atomic order translates in real space by an analysis in terms of cluster hierarchy. This hierarchy is quite close to that encountered in chapter 5 for the flattened polytopes. The analogy is in fact quite deep, but requires to be displayed, to study a regular lattice in eight dimensions, the E8 lattice, which gives upon suitable mapping a quasicrystal in four dimensions with the $\{3, 3, 5\}$ symmetry. Geometry and number theory enter the game in a rather technical way, the principal results being given in appendix A9. The interested reader should look at the original papers (Elser and Sloane 1987, Sadoc and Mosseri 1993).

Leaving the 'perfect' quasiperiodic structures, we shall then focus on questions related to the configurational entropy in the so-called 'random tiling' models, obtained by disordering the perfect quasiperiodic tilings through iterated local transformations called 'elementary flips'. This description bears some relation to that in chapter 4 in the context of the corrugated space model. In addition, we shall argue on the possible analogies between the kinetic properties of quasicrystals and glasses.

8.2 Hierarchical clusters in quasicrystals

The best known quasiperiodic tilings, pentagonal and octagonal in two dimensions and icosahedral in three dimensions, all share the property of self-similarity. As a consequence, the limited number of local environments, which we can easily enumerate, reproduce at all scales and are interlaced in a hierarchical way. This property remains when one shifts from tilings to atomic structures (for instance by decorating the tiles) and one expects then to find hierarchical clusters of atoms. One can usually distinguish two principal families of quasicrystalline structures, either of the AlLiCu type, based on compact polytetrahedral packings similar to those found in the Bergman phase (see chapter 7), or of the AlCuFe and AlPdMn type, whose local order is best described in terms of Mackay icosahedra (see chapter 3), which are more or less perfect (Janot 1992). The suggestion that the structure of these phases should be based on such a hierarchy of clusters was made from the start, inspired by the analysis in terms of polyhedral shells of the neighbouring large cell crystals, of the Frank–Kasper type (Audier and Guyot 1988). It is therefore interesting to recall the analogy, raised a few years before, and described at length here, in particular in chapter 7, between Frank–Kasper-like large cell crystals and systems with frustrated icosahedral order. The hierarchical cluster structures, proposed to describe quasicrystals, are probably very close to those obtained from the hierarchical polytopes described in chapter 5.

A further step has been taken by Janot and de Boissieu (1994) who propose that this hierarchical structuring is important not only at a descriptive level, but also for explaining the stability of these phases. They use a simplified model for the electronic structures of quasicrystals, with a sequential filling of energy levels (calculated within a jellium-type approach) associated with the several steps of aggregation. This interesting approach can nevertheless be criticized for its 'molecular crystal' picture of the quasicrystalline state. Indeed, such a model supposes a depletion of electrons on the atoms which link the different clusters, which remains to be proved, while nothing would prevent a similar description, but based on hierarchical clusters which interpenetrate, making then the distinction more difficult between internal and linking atoms.

8.3 Random tilings

The origin of quasicrystal stability is far from understood, and this is not really a surprise owing to the complexity of their phase diagram. Different models, of classical as well as of quantum nature, are in competition. The simplest idea, suggested by the existence of 'matching rules' in the perfect tilings, consists of looking for a classical hamiltonian built on pair-wise interactions, which would have a quasiperiodic ground state. Although this has been shown to be possible, but with *ad hoc* potentials, no convincing model, sufficiently generic, has yet been developed. A main difficulty is that pair potentials, of limited range, cannot discriminate between almost equivalent configurations at a medium range, whose presence can break the tiling perfection.

This degeneracy is at the heart of the so-called 'random tiling model' (Elser 1985), an approach which gives the particular positional and orientational order of quasicrystals an entropic origin. As will be explained below, one considers ensembles of random tilings, associated with a discrete configuration space, the latter being endowed with a minimal distance which corresponds to the simplest topological rearrangements, called localized phasons, or 'elementary flips'. These tilings are derived from canonical tilings, obtained by projection from a hypercubic tiling in high dimension, and are therefore made of rhombi in two dimensions and rhombohedra in three dimensions.

We today believe that, as far as quasicrystal stability is concerned, the main, first order, cause is to be found at the level of their electronic structure, in the Hume-Rothery-like framework (Friedel and Denoyer 1987), at least for the best quasicrystals, which would then favour the more ordered quasiperiodic structures upon their disordered variants.

Studying random tiling models is nevertheless interesting for several reasons. Firstly, the calculated entropies are indeed high, and one expects that these models may play a role at high enough temperature (although one should not forget that from tilings to structural models, the atomic decoration significantly decreases this configuration entropy).

In addition, from a statistical mechanics point of view, random tilings prove to be a very interesting class of models, some being exactly solvable.

Finally, the atomic diffusion associated with phason jumps could be enhanced at high temperature. A transition between a temperature range where phason jumps are 'frozen' and a higher temperature regime where diffusion paths percolate in the material has been proposed (Kalugin and Katz 1993). An analogy has been suggested, in reference to the classical free volume model with the glass transition phenomenon (Mosseri 1994). This analogy could even be extended more generally to more quasicrystal properties, this point being discussed later.

We do not aim here at an exhaustive description of these models. In a first step, we shall restrict ourselves to the so-called 'maximal random tilings', where all tilings which can be transformed into one another by a sequence of elementary flips have the same energy – it corresponds to the infinite temperature limit of the more general case. We then discuss some ideas to discriminate among these tilings, through specific tiling energy models.

We first describe the one-dimensional case, for which entropy is easily calculated, and then consider a first non-trivial example of a random tiling, the rhombus hexagonal tiling, where each realization is in one-to-one correspondence with one ground state of the antiferromagnet Ising model on a triangular lattice, whose solution goes back to Wannier (1950). It is interesting to consider here some particular boundary conditions, first studied by Elser (1984). He showed that, with a hexagonal boundary, this problem is formally equivalent to the old combinatorial problem of 'planar partitions', extensively studied in the past by MacMahon (1916). A simple expression for the configuration entropy can be derived. This analysis can be extended, by defining generalized planar partitions, to rhombus tilings of higher codimension (Mosseri and Bailly 1993). The octagonal case, obtained by projection from four to two dimensions, will be presented. Finally, the simplest example of a tri-dimensional tiling (that with codimension 1) will be described (Mosseri *et al.* 1993). The interested reader can find a more extended version in the lecture given at the Aussois winter school on quasicrystals (Mosseri 1994).

8.4 Random tilings in one dimension

Let us call $W^{2,1}_{k_1,k_2}$ the number of ways to tile a line with two types of segment (with k_1 and k_2 units of each respectively). This number is simply given by the binomial coefficient

$$W^{2,1}_{k_1,k_2} = \frac{(k_1 + k_2)!}{k_1! k_2!} = \binom{k_1 + k_2}{k_1} \qquad (8.4.1)$$

If the number of segments is increased, while keeping constant the ratio k_1 to k_2, the following expression is found for the entropy S per site:

$$S^{2,1}(x_1, x_2) = -(x_1 \log x_1 + x_2 \log x_2) \quad \text{with} \quad x_i = \frac{k_i}{k_1 + k_2} \qquad (8.4.2)$$

As noted by DiVincenzo (1988a), it corresponds, in the 'cut and project' algorithm, to considering a random walk in the perpendicular space direction (the 'physical space' being the time direction), the Gaussian nature of this phenomenon arising naturally from the asymptotic behaviour of the binomial coefficients. It is useful, for the remainder, to note that each configuration can be put in correspondence with a one-dimensional partition, e.g. an ordered set of k_1 integers $\{v_l\}$, in the range between zero and k_2, and such that $v_l \leqslant v_{l+1}$ (see figure 8.1). In the notation for S and W, 2,1 means that the one-dimensional tiling is 'lifted' as a path in two dimensions.

In the random tiling model language, one defines the phason strain E:

$$E = \frac{k_1 - k_2}{k_1 + k_2} \qquad (8.4.3)$$

Walk on a square lattice

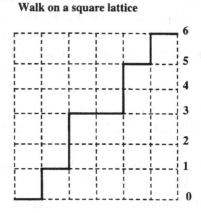

Linear Partition

$$0 \leq 1 \ \leq \ 3 \leq \ 3 \ \leq \ 5 \leq 6$$

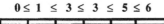

Fig. 8.1. An example of a linear partition and the associated path in the square lattice.

and the entropy, expanded near its maximum (for $x = 0.5$) reads

$$S^{2,1}(E) = S^{2,1}(0) - \frac{1}{2} KE^2 + \frac{1}{12} E^4 + \cdots \qquad (8.4.4)$$

where the phason elastic constant K takes the value 1. This quadratic behaviour, which is trivial here, is a central hypothesis of the model in higher
dimensions, which allows us to treat phasons and phonons almost on the same
footing.

It is easy to generalize the above case to linear tilings containing d different
types of segment. One then needs to embed the chain in a d-dimensional
hypercubic lattice. Counting the configurations uses standard 'multinomial'
coefficients and the entropy reads

$$S^{d,1}(x_1, \ldots, x_d) = - \sum_{i=1}^{d} x_i \log x_i \quad \text{with} \quad x_i = \frac{k_i}{\sum_{i=1}^{n} k_i} \qquad (8.4.5)$$

8.5 Two-dimensional tilings

8.5.1 The codimension 1 case: the rhombus hexagonal tiling

This tiling is obtained by selecting a 'connected membrane' built with the
square faces of a cubic tiling. This membrane is called 'directed', to emphasize
that it leads to a perfect tiling (without overlaps or holes), when projected
orthogonally on the $(1,1,1)$ plane, with three types of rhomb equivalents under
a six-fold rotation. The enumeration of all configurations is much more
complicated than in one dimension, but can be mapped, as said above, to
already solved models (Blöte and Hilhorst 1982). If the tiling is constrained
with a hexagonal boundary, one can use the even older solution of the planar
partition problem. Note that this model was used to estimate the interfacial
tension (supposed to be mainly of entropic origin) between a solid and an
undercooled liquid (Spaepen and Meyer 1976).

A 'restricted' planar partition amounts to assigning integers, in the range
from zero to p, to cells of a square tiling, with the additional constraint of
'descending order', which means that the number assigned to a square cell is
greater than or equal to the numbers associated with cells immediately on its
right and below. The lift as a discrete membrane in the cubic tiling is done by
considering this number as the 'altitude' of this square face. The two other
types of (vertical) face are then added in order to ensure the corrugated surface
connectivity. See figure 8.2.

Upon mapping along the $(1,1,1)$ direction, one gets a rhombus tiling inside a
hexagonal boundary of integral sides k_1, k_2 and p. It is easy to see that, by

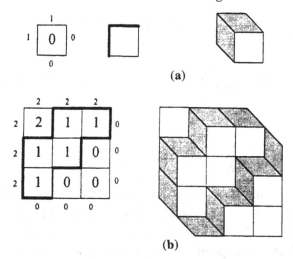

Fig. 8.2. The partition problem in the case $3 \rightarrow 2$. Integers, in the range from zero to k_3, are assigned to the square tiles with a descending order constraint. (a) A configuration in the case (1,1,1). (b) A configuration in the case (3,3,2).

symmetry, p and k_i play the same role, and we therefore rename p as k_3. Note that this hexagonal boundary corresponds, in the cut and project framework, to suitably selecting points inside a parallelepiped of sides k_1, k_2 and k_3.

A closed expression for the number of planar partitions (and therefore of codimension 1 rhombus tilings) was given by MacMahon. It is interesting to rewrite it in terms of what was called 'generalized factorials' (Mosseri *et al.* 1993), defined as follows:

$$k!^{[p]} = \prod_{l=1}^{k} l!^{[p-1]} \quad \text{with} \quad k!^{[0]} = k \quad \text{and} \quad k!^{[1]} = k! \tag{8.5.1}$$

These generalized factorials behave asymptotically as

$$k!^{[p]} \rightarrow k^{k^p/p!} \quad \text{when} \quad k \rightarrow \infty \tag{8.5.2}$$

With this notation, the classical MacMahon formula reads:

$$W^{3,2}_{k_1,k_2,k_3} = \frac{(k_1 + k_2 + k_3 - 1)!^{[2]}(k_1 - 1)!^{[2]}(k_2 - 1)!^{[2]}(k_3 - 1)!^{[2]}}{(k_1 + k_2 - 1)!^{[2]}(k_1 + k_3 - 1)!^{[2]}(k_2 + k_3 - 1)!^{[2]}} \tag{8.5.3}$$

where W counts the number of partitions. One can easily derive a formula for the configuration entropy (per tile), by taking k_i infinite while keeping their ratio constant, and using the above asymptotic behaviour for the generalized factorials:

$$S^{3,2}(x_1, x_2, x_3) = \frac{1}{2X} \left(\sum_{i=1}^{3} (x_i^2 \log x_i - (1 - x_i)^2 \log(1 - x_i)) \right)$$

with

$$x_i = \frac{k_i}{\sum_{i=1}^{3} k_i} \quad \text{and} \quad X = x_1 x_2 + x_1 x_3 + x_2 x_3 \qquad (8.5.4)$$

This formula was previously derived by Elser (1984). The total number of rhombs is $k_1 k_2 + k_1 k_3 + k_2 k_3$ where $k_i k_j$ is the number of rhombs of the type defined by the edges (i, j). The entropy reaches its maximum (0.2616) when each $x_i = 1/3$, which corresponds to an average orientation of the physical space parallel to the (1,1,1) reticular plane, and to an equal proportion for the three types of rhomb. Note that the entropy is less than that obtained with free boundary conditions. Elser gave a clue to explain the origin of that difference, in terms of an entropy gradient from the boundary to the bulk, and a connection between the two cases (free and fixed boundary) was studied in more detail recently (Destainville *et al.* 1995).

8.5.2 Two-dimensional rhombus tilings of codimension 2

This case is important since it corresponds to the eight-fold symmetric experimentally observed quasicrystals (Wang and Kuo 1988). A perfect octagonal tiling had already been introduced by Beenker (1982), and was in fact anticipated several years before by Amman (see Grünbaum and Shepard 1987, page 556). It is generated, in the cut and project scheme, by selecting sites inside a hypercubic four-dimensional lattice, and projecting onto a suitably oriented two-plane. It contains two types of tile, a square and a rhombus with two 45° angles. No exact solution is yet known for the configuration entropy in this family of tilings, but numerical results, obtained from transfer matrix calculations, are available (Li *et al.* 1992).

8.5.3 Generalized planar partitions

We consider the set of tilings, by squares and 45° rhombi, inside an octagonal boundary. This boundary is the 'shadow', in physical two-dimensional space, of the big parallelotope inside which the points are selected in four dimensions. The enumerating question is equivalent to a generalized partition problem

(Mosseri and Bailly 1993), defined as follows. We first generate standard planar partitions as in §8.5.1. To each such partition, there corresponds a rhombus tiling with hexagonal boundary of integral size (k_1, k_2, k_3). Then, on each such tiling, we define a new partition problem by assigning to each rhomb an integer in the range between zero and k_4 (with a suitable descending order, see figure 8.3).

The hexagonal boundary is transformed into an octagonal one of sides k_1, k_2, k_3 and k_4. The full set of allowed configurations is generated. Note that, as opposed to the codimension 1 case, the boundary may display different shapes (three at most, modulo simple symmetries) according to the order among the k_i. The total number of configurations is obtained, for a given quadruplet

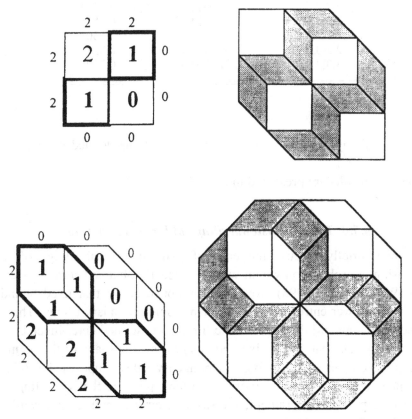

Fig. 8.3. Generalized partition problem in the case $4 \rightarrow 2$. First, a configuration of the $3 \rightarrow 2$ case is generated. Integral numbers between zero and k_4 are then assigned to all rhombic tiles with prescribed descending order. Zero and k_4 values are assigned to the boundary. Edges separating neighbouring cells whose assigned values differ are 'enlarged' into new pseudo-linear arrangements of rhombs, which amounts in fact to associating these integral values to 'altitudes' in the fourth dimension.

Table 8.1. *Number of configurations and entropy per tile as a function of the octagonal boundary size*

k_1, k_2, k_3, k_4	No. of configurations	Entropy/tile
1,1,1,1	8	0.3466
1,1,1,2	20	0.3329
1,1,1,3	40	0.3074
1,1,1,4	70	0.2832
1,1,2,2	75	0.2878
1,2,1,2	76	0.2887
1,2,2,2	480	0.3430
2,2,2,2	5383	0.3580
2,2,2,3	39 202	0.3525
2,2,2,4	213 298	0.3408
2,2,2,5	937 362	0.3274
2,2,3,3	442 922	0.3514
1,3,3,3	146 724	0.3305
2,3,3,3	8 473 872	0.3545
3,3,3,3	271 776 385	0.3596

(k_1, k_2, k_3, k_4), by summing the number of partitions defined on each codimension 1 solution.

Numerical results are presented in table 8.1.

8.5.4 Two-dimensional tilings of large codimension

Let us now briefly discuss the case of two-dimensional tilings of large codimension, a problem on which we have recently obtained some interesting results (Widom *et al.* 1998). We have mentioned above that fixed boundary tilings have a lower entropy than free (or periodic) boundary tilings. The very existence of a standard thermodynamic limit is questioned here. Indeed, a first non-trivial point is that, in the fixed boundary case, the shape of the boundary (the respective sizes of the polygonal boundary sides) imposes the different tiles stoichiometry. A second, more important, point is that the tilings display an entropy density gradient, increasing from the boundary towards the centre.

For large codimension tilings, we have shown that a standard thermodynamic limit is recovered. Even more, the tiling entropy does not depend anymore on the precise proportion of the different types of tiles, but depends only on the codimension when the latter goes to infinity. We have shown that, in this limit, the entropy per tile is bounded above by log(2). This is clearly supported by some exact enumerations (up to the $10 \to 2$ case), and extended simulations (up to $d \simeq 100$).

8.5.5 *Remarks on the random tiling geometry*

An important hypothesis of the maximal random tiling model is that all configurations share the same energy, a kind of infinite temperature approximation. To get a more realistic representation, we need to be able to distinguish between different configurations. As an example, it has been proved that one can always transform a quasiperiodic tiling into a multi-twinned crystal (with crystallites of any size) through sequences of elementary flips (Mosseri 1993). How can we describe these two extreme cases in a unified approach, which clearly marks their differences? We give here some suggestions in this direction, firstly in terms of the disclination set associated to the tilings. This point of view is reminiscent of the so-called 'corrugated space approach' (Gaspard *et al.* 1982), presented in chapter 4. This description of complex structures in terms of corrugated hyper-surfaces (of variable local curvature but flat on the average), bears some resemblance to the idea of lifting into a high-dimensional space which is at the heart of the cut and project method, with the difference that only codimension 1 surfaces were thought of in the former case.

Any rhombus tiling has an average coordination number of four, a result that can be easily proved with the help of the Euler–Poincaré relation. Four-fold coordinated sites play the role of neutral charges (see chapter 4), those of higher (lower) coordination being on negative (positive) disclinations. A pair of neighbouring sites of coordination three and five form a dislocation. Lifted in the higher-dimensional space, this defect appears as a step corner on the surface (figure 8.4). Note that the only sites which can be flipped are on positive disclinations. If one compares the set of disclinations in quasiperiodic and multi-twinned tilings (figure 8.4), they appear clearly more complex in the former case, where the average distance between defects is smaller. Note in addition the screening effect between disclinations of opposite signs, which is in fact related to the very existence of the tiling (the Gaussian curvature of the surface must vanish on the average on small regions in order to avoid some tiles overlapping after mapping). Disclinations interact in two dimensions like electric charges, with logarithmically varying potentials.

It would be interesting to study, from this point of view, the ground state of sets of charges confined in two dimensions inside regular polygons of size $2n$

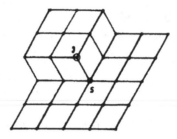

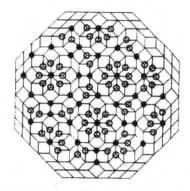
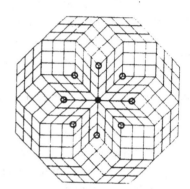

Fig. 8.4. Defects associated with rhombic tilings. Upper drawing: a pair of positive–negative disclinations forming an isolated dislocation, which appears as a step corner in the embedding space. Note that the only sites which can be flipped are on positive disclinations. Left: set of disclination defects in an almost perfect octagonal tiling. Right: set of disclination defects in a micro-twinned tiling.

(starting with $n = 4$), with an excess of positive charges inside and the possibility of generating at will pairs of positive and negative charges.

A second way to introduce energy considerations in this problem involves going back to the analogy between the tilings and the antiferromagnetic Ising model in the triangular lattice. The ground state degeneracy is broken if one adds a second neighbour interaction, weak enough that the low energy configurations remain among the degenerate set of ground state solutions of the former problem. Indeed, only these solutions can be unambiguously associated with tilings. In these conditions, one can show that a second neighbour antiferromagnetic interaction selects the multi-twinned tilings, while a ferromagnetic case favours the state corresponding to a membrane (in three dimensions, before projection), which is as flat as possible. This solution, crystalline in this case, is analogous to the most symmetrical quasicrystalline tilings in higher codimension. This approach is quite promising, and will be discussed again in §8.7.

8.6 Three-dimensional rhombohedral tilings

8.6.1 The codimension 1 case

The problem consists of enumerating rhombohedral random tilings in R^3, obtained by projecting a corrugated three-dimensional hyper-surface (a 3-surface), built on a four-dimensional hypercubic lattice, and attached in that space to a hyper-rhombohedron of size k_1, k_2, k_3, k_4.

Such a tiling is generically made of four different rhombohedra and bounded, in R^3, by a rhombic dodecahedron (the shadow of the projected hyper-rhombohedron). This is again equivalent to a combinatorial problem, the so-called 'solid partitions' problem, for which no exact enumerative formula is known. Nevertheless, an approximate solution was proposed, which gives values quite close to the exact numerical ones for small systems (Mosseri *et al.* 1993):

$$W^{4,3}_{k_1,k_2,k_3,k_4} = \frac{(k_1 + k_2 + k_3 + k_4 - 2)!^{[3]} \prod_{\langle l,m \rangle}(k_l + k_m - 2)!^{[3]}}{\prod_{\langle l,m,n \rangle}(k_l + k_m + k_n - 2)!^{[3]} \prod_{l=1}^{4}(k_l - 2)!^{[3]}} \tag{8.6.1}$$

where $\langle l, m \rangle$ means a sum over the six pairs of indices and $\langle l, m, n \rangle$ a sum over the four triplets.

Table 8.2 compares the values obtained from the approximate formula to the exact ones, obtained numerically, for different boundary conditions. The last column gives the difference between the exact and the approximated entropy, in the form of $D = 10^4/N \log(V/W)$, where N is the number of rhombohedra. The last exact value, in the case (4,4,4,4), was obtained thanks to a new approach to the problem, which transforms it into the counting of integral points inside a convex polytope in very high dimension (Destainville *et al.* 1995), which is considerably quicker to enumerate.

Going back to the approximate formula, one remarks that it always gives values lower than the exact ones. We believe that they form the first (and asymptotically dominant) term of a development for the exact value. We can then take the limit of the approximate formula to propose an ansatz for the entropy

$$S^{4,3}(x_1, x_2, x_3, x_4) = \left(-\sum_{i=1}^{4} x_i^3 \log x_i + \sum_{i,j}(x_i + x_j)^3 \log(x_i + x_j) \right.$$

$$\left. - \sum_{i=1}^{4}(1 - x_i)^3 \log(1 - x_i) \right) \bigg/ 6X$$

Table 8.2. *Number of configurations in rhombohedral tilings of codimension 1: approximate formula W, exact enumeration V and difference (multiplied by 10^4) for the entropy per tile, D, for different limit conditions. In the last line:*
$$W = 523\,207\,623\,809\,678\,438 \text{ and } V = 728\,321\,630\,294\,356\,72$$

k_1, k_2, k_3, k_4	W	V	D
1,1,1,1	2	2	0
2,2,2,2	166.66	168	2.49
2,2,2,3	875	887	3.096
2,2,2,4	3430	3490	3.097
2,2,2,5	10976	11 196	2.92
2,2,2,6	30 240	30 900	2.7
2,2,3,3	575	8790	4.13
2,2,3,4	57 624	59 542	4.31
2,2,3,5	296 352	307 960	4.18
3,3,3,3	16 263 797.76	17 792 748	8.32
4,4,4,4	see caption	see caption	14.1

with

$$x_i = \frac{k_i}{\sum_{i=1}^{4} k_i} \quad \text{and} \quad X = x_1 x_2 x_3 + x_1 x_2 x_4 + x_1 x_3 x_4 + x_2 x_3 x_4 \tag{8.6.2}$$

The validity of this expression, which would give a new solvable model in three dimensions, is difficult to establish. Note that the exact entropy, in the diagonal case and for increasing values of k_i, shows a slow decrease towards the value given by the ansatz:

$$S(1/4,\ 1/4,\ 1/4,\ 1/4) = 1/6(44 \log 2 - 27 \log 3) \approx 0.139 \tag{8.6.3}$$

8.6.2 The icosahedral case in three dimensions

As in the two-dimensional case, it is possible to describe a rhombohedral tiling in three dimensions of higher codimension by defining generalized solid partitions (constructed by an iterative process). The most interesting case is that of codimension 3, for which the entropy is maximal when the average orientation of the (three-dimensional) surface in R^6 coincides with that giving the icosahedral quasicrystals. This amounts to iterating twice on the above solid partition.

8.6.3 Geometrical considerations: disclination lines

As in two dimensions, it is possible to analyse these configurations in terms of defects, here disclination lines of both signs, according to the number of rhombohedra sharing the edges, neutral edges sharing four rhombohedra. Sites which can rearrange by 'flip' are nodes in the network of positive disclinations where four defects meet.

Note the following point, interesting in the context *à la* Regge (see chapter 4) of discrete approximations of curved manifolds. We have seen that in three dimensions, as opposed to the two-dimensional case, approximations must be made in order that curvature be coded, in an imperfect manner, in the angular deficit along edges. Consider now the above 3-surfaces, embedded in the higher space. For a fixed boundary condition, all the (very numerous) configurations compatible with that condition, can be transformed into each other with sequences of elementary flips. But these 'flips' do not change the total length of edges sharing three rhombohedra. In the Regge sense, the curvature content of these 3-surfaces does not change, being fixed by the boundary conditions.

8.7 Glass-like properties in quasicrystals

Despite the many successes obtained in the structural description of quasicrystals, the debate about the best way to characterize these materials is still open. Should we regard them as being close to crystals, owing to their long range positional and orientational order? Should we better, stressing their physical properties and their local order, consider them as being closer to glasses? Let us discuss now the possibility, and the reasons, of why we should characterize these properties as being of the vitreous kind (Mosseri and Sadoc 1995). One should at first sight be surprised by this way of describing a rather well characterized material (the quasicrystal) in terms of a less well understood one (the glass). It might indeed prove true that such an analogy would be more helpful for the physics of glasses, but we also feel that it will be useful in the quasicrystal field.

Our main point is to consider the energy landscape in the configuration space of these materials, i.e. the space where each point represents a given structural realization. Simple crystals are believed to correspond to well distinguished deep minima, while one expects a much more complex and hierarchical structure in the case of glass (and in spin glasses), with a highly degenerate ground state and a high density of tunnel states. We suggest here that quasicrystals also present such a hierarchical energy landscape.

Compared to crystals, quasicrystals possess the very important additional phason degree of freedom, stressing here the inhomogeneous, localized phasons which carry the complexity and the configuration entropy of these systems. In the 'maximal random tiling scenario' considered above, all the structures, differing by a sequence of flips, are iso-energetic. In addition, the configuration space is believed to be 'ergodic', in the sense that two tilings can be transformed into each other through a sequence of flips (note, however, that this ergodic property is not yet proved for three-dimensional tilings). At this level, we are therefore very far from glasses, where a broken ergodicity phenomenon is expected. But, as already discussed above, more realistic models should discriminate between these different tilings. Consider for example the decoration–deflation process of generating quasiperiodic tilings. To a perfect tiling there corresponds a very precise sequence of operations. In such a sequence, we say that a tile P 'covers' a given region, to the order n, if all the tiles in this region come from the tile P after n decoration steps. Suppose now that flips are randomly operated during the iterations, followed infinitely by decorations. If a flip occurs, and if no other flip is ever made in the region concerned by this flip, a perfect quasiperiodic tiling is therefore generated in that region. It is, however, separated from the remainder of the tiling (also made of limited regions of perfect tiling but differently oriented) by sites and tiles forming what can be called 'antiphase boundaries'. But it is possible to generate a large number of such structures, by randomly mixing flips and deflation, which will display different scales of order, but very close energies, which is in first order proportional to the length (or area in three dimensions) of these boundaries. A structure 'quenched' from a high temperature random phase, should reach quite rapidly one of these structures (rather than the perfect quasiperiodic one without defects). The growth of the ordered phase (coarsening) requires then that the defect boundaries be eliminated. If this implies moving an increasing number of sites, over increasing distances, the free energy barrier heights will grow with the ordered region size, and one faces a situation close to that invoked for glasses (Shore et al. 1992), with diverging energy barriers, a highly complex energy landscape, and broken ergodicity. One expects very slow coarsening, even logarithmic with time. Such a scenario could be numerically tested, with for example the simple energy model discussed above (§6.5).

Note that this point of view is close to that already discussed in chapter 5 in order to model the glass transition by disordering the sequence of hierarchical polytopes. In addition, in the present model, which is only conjectured here, the perfect quasicrystal corresponds to an 'ideal' structure for glasses and undercooled liquid, a point of view which was already present, or implicit, in early studies of quasicrystals.

Up till now, we have only considered global properties of the random tilings. But the flips may also trigger site autodiffusion, particularly in codimensions greater than one, where subtle topological effects appear, which were originally underlined in the form of non-trivial site permutations (Frenkel *et al.* 1986). The role of atomic displacements was then discussed in several works (Duneau and Oguey 1990, Coddens and Launois 1991). A mechanism for the transition between quasiperiodic structures and multi-twinned crystals was demonstrated, which integrates site autodiffusion (Mosseri 1993). This kind of diffusion mechanism was rationalized by Kalugin and Katz (1993), who in addition proposed a scenario where long range diffusion requires that infinite diffusion paths throw open thermal activation.

Such a scenario recalls some aspects of the older 'free volume' model of the glass transition, and leads us to consider the possibility of glass-like kinetic behaviour in quasicrystals. The glass transition is a universal phenomenon, found in many undercooled liquids, where macroscopic physical properties vary, while no big changes are observed at the level of atomic structure. One usually considers that atomic diffusion is frozen near T_g, the glass transition temperature (which is in fact a temperature range). In the free volume model, each atom (or molecule) moves in a cage bounded by its first neighbours. In the undercooled liquid, there are many configurations where an atom can jump towards a neighbouring cage with a similar local environment. When the temperature decreases, these possibilities eventually vanish, diffusive motions almost stop and only vibration-like motions survive (inside cages). The material is frozen in a vitreous state. According to Cohen and Grest (1979), the glass transition coincides with the free volume percolation threshold. Now, important ingredients in this description are found in random tilings. Consider first a site which can be flipped. The two local configurations provide an ideal pattern corresponding to the free volume cages, the flip giving precisely the jump between the two close configurations (this is even clearer if one looks at the atomic decorated tiles). As said above, a percolation-like phenomenon has been suggested for the site autodiffusion in the random tilings. The question of whether such a concept of free 'movement' (which was implicit in the free volume model) could explain the spectacular fragile–ductile transition in quasicrystals (through dislocation unpinning) is still being debated.

A1

Spaces with constant curvature

A1.1 The three geometries

Spherical, Euclidean and hyperbolic geometries are the three types of homogeneous geometries on which a two-dimensional manifold can be locally modelled. They have a constant Gaussian curvature which is positive in spherical space, null in Euclidean space and negative in hyperbolic space. In three dimensions these three geometries have an analogue, but there are five other homogeneous geometries (Thurston 1982, Scott 1983). In this section, we briefly present Euclidean and hyperbolic geometries, while the next section is devoted to spherical spaces. Section A1.3 is concerned with the more exotic cylindrical spaces and §A1.4 presents the different notions of curvature.

A1.1.1 Euclidean geometry

Two- and three-dimensional Euclidean spaces do not require any long description. The main transformations which let the space be globally invariant are translations (which let no points be invariant), rotations (with one point – in two dimensions – and one line – in three dimensions – kept invariant) and mirror inversion (with one line – in two dimensions – and one plane – in three dimensions – kept invariant). Note that any rotation can be viewed as the combination of two mirror inversions, a property which will be illustrated later with the groups generated by reflections, like in a kaleidoscope (appendix A4).

A1.1.2 Hyperbolic geometry

Hyperbolic geometry differs from the Euclidean geometry by the basic axiom concerning parallelism. Now 'through a point not on a line, there can be more

214

than one line parallel to the given line'. This is quite unusual and we shall now give a quick description of what has also been called the Lobachevski–Bolyai geometry, with reference to its two independent discoverers.

The hyperbolic plane H^2 cannot be embedded as a whole in R^3. It is however possible to give conformal representations in R^2 (which preserves angles): let us study one such model, introduced by Poincaré, known as the unit disc model. The points of the hyperbolic plane are represented by points which belong to the interior of a fixed unit disc limited by a circle, called the absolute, whose points correspond to the points at infinity (figures A1.1–A1.2).

The geodesic lines of H^2 are circles orthogonal to the absolute and reflections about these lines are replaced by 'inversion' with respect to these circles. If the point coordinates are given in terms of a complex number z, the group of isometries of H^2 is the group of substitutions

$$z' = \frac{az + b}{\overline{b}z + \overline{a}} \tag{A1.1}$$

with $a\overline{a} - b\overline{b} = 1$ and a, b, z, $z' \in C$ in which one includes the reflection in the imaginary axis:

$$z' = -\overline{z}$$

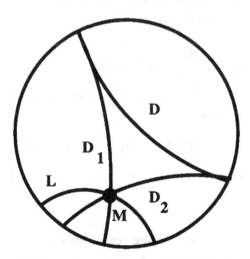

Fig. A1.1. The Poincaré disc: points of the hyperbolic plane are represented inside a disc of unit radius. The limiting circle, the 'absolute', corresponds to points at infinite distance. Geodesic lines of H^2 (some of which are represented here) are given by circles orthogonal to the absolute.

Fig. A1.2. An artist's view of the Poincaré disc. M.C. Escher 'Circle limit IV', 1996. Cordon Art – Baarn – Holland. All rights reserved.

H^2 is a manifold of constant negative curvature, and the square of the line element reads (with a unit curvature radius):

$$\mathrm{d}s^2 = \frac{4}{(1 - |z|^2)^2} \, \mathrm{d}z \, \mathrm{d}\bar{z} \qquad\qquad (\mathrm{A1.2})$$

The three-dimensional hyperbolic space H^3 can be represented similarly as the points inside the Euclidean ball:

$$H^3 = \{(x, y, z); \ (x_1^2 + x_2^2 + x_3^2) < 1\}$$

with an appropriate metric which we shall not describe further.

A1.2 Spherical spaces

A1.2.1 S^2, the 2-sphere

A two-dimensional spherical space can be embedded in three-dimensional Euclidean space in the form of the usual sphere, denoted S^2, of radius r and constant Gaussian curvature $\kappa = 1/r^2$ (see §A1.4 about the notion of intrinsic curvature). This intrinsic curvature can be verified by direct measurements on the surface itself: for instance the area of a geodesic triangle is $S = (2\pi - A - B - C)/\kappa$ where A, B and C are the angles of the triangle.

A sphere S^2 is currently defined in an orthonormal coordinate system as $x_1^2 + x_2^2 + x_3^2 = r^2$ but displacements on the surface are better defined by using the two angles θ and ϕ of the spherical (or polar) coordinate system:

$$x_1 = r \cos \theta$$

$$x_2 = r \cos \phi \sin \theta \qquad (A1.3)$$

$$x_3 = r \sin \phi \sin \theta$$

with $\theta \in [0, \pi]$ and $\phi \in [0, 2\pi[$. The two parameters θ and ϕ define an intrinsic representation of S^2. If lines of constant θ and ϕ are drawn, they form a mesh of the surface, but with two singularities located at the two poles. Geodesic lines are great circles, any other circle on S^2 being smaller. Two geodesics have two common points diametrically opposite on the sphere.

A1.2.2 S^3 or the 3-sphere: spherical coordinates

Adding one dimension, we get the hypersphere S^3, which is a three-dimensional space of constant, homogeneous and positive curvature. This space is usually described by four coordinates in R^4 as

$$x_0^2 + x_1^2 + x_2^2 + x_3^2 = r^2$$

However, some important aspects of this space are better seen by changing the point of view, and using two other coordinate systems, the toroidal (see below) and spherical systems. In the latter case, the coordinates of one point are expressed as functions of the radius r and of three suitably defined angles θ, ϕ and ω. This is indeed a natural extension of the polar system used above for S^2.

The relation to the orthogonal coordinates in R^4 is given by:

$$x_0 = r \cos \omega$$

$$x_1 = r \cos \theta \sin \omega$$

$$x_2 = r \cos \phi \sin \theta \sin \omega \qquad (A1.4)$$

$$x = r \sin \phi \sin \theta \sin \omega$$

with $\omega, \theta \in [0, \pi]$ and $\phi \in [0, 2\pi[$. Note at once that, fixing $x_0 = 0$ (e.g.

$\omega = \pi/2$), the above polar system of S^2 is recovered. Moreover, one clearly sees that surfaces at constant ω are spheres S^2 with radii $r \sin \omega$. Special among them is the great sphere S^2, obtained for $\omega = \pi/2$, which is equidistant from the two poles and separates S^3 into two equivalent sub-spaces. So, embedded in S^3, we can define great 2-spheres which divide the space into two half hyperspheres, exactly like an equator divides S^2 into two north and south hemispheres. Varying θ and ϕ by $d\theta$ and $d\phi$ then leads to the orthogonal displacements $r \sin \omega \, d\omega$ and $r \sin \theta \sin \omega \, d\phi$ on the sphere S^2, while changing ω by $d\omega$ leads to a displacement $r \, d\omega$, normal to the sphere S^2.

While speaking of areas and volumes, it is worth introducing a new term, the *d*-volume, in order to avoid confusion when changing dimensions. A 2-volume is the usual area, a 3-volume is either the standard volume (interior) of a 2-sphere, or the (hyper)surface of a 3-sphere. The element of 2-volume (area) on the 2-sphere defined by a constant ω therefore reads $da = r^2 \sin \theta \sin^2 \omega \, d\theta \, d\phi$ and the element of 3-volume in S^3 is $dv = r^3 \sin \theta \sin^2 \omega \, d\theta \, d\phi \, d\omega$ so that the 2-volume of a small sphere in S^3 is $A = 4\pi r^2 \sin^2 \omega$ and its 3-volume $V = 2\pi r^3(\omega - (\sin^2 \omega)/2)$. Note that some care should be taken when speaking of the 3-volume of S^2 in S^3, since a sphere S^2 separates S^3 into two parts, whose 3-volumes add to the full S^3 3-volume: $V = 2\pi^2 r^3$. Here, by convention, the calculated 3-volume corresponds to that part which contains the 'north pole' $\omega = 0$.

In particular, one notes that the great sphere S^2, with $\omega = \pi/2$ and radius r has a 2-volume $4\pi r^2$ and a 3-volume $\pi^2 r^3$, the latter being exactly half of the S^3 3-volume. Finally, note that great circles can be drawn on a great sphere: such great circles are geodesic lines of the three-dimensional spherical space. See figure A1.3.

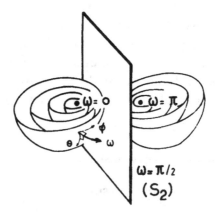

Fig. A1.3. Stereographic projection in R^3 of a family of spheres with $\omega =$ constant in S^3; here, the great sphere S^2 corresponding to $\omega = \pi/2$ contains the projection pole. θ and ϕ are variables lying on the surfaces of the 2-spheres.

A1.2.3 Toroidal coordinates

It is also possible to define toroidal coordinates as follows:

$$x_0 = r \cos \phi \cos \theta$$

$$x_1 = r \cos \phi \sin \theta$$

$$x_2 = r \sin \phi \cos \omega \tag{A1.5}$$

$$x_3 = r \sin \phi \sin \omega$$

with ω, $\theta \in [0, 2\pi[$ and $\phi \in [0, \pi/2]$. In this system, surfaces with constant ϕ are tori organized around their symmetry axes which are the two interlaced great circles, $x_0^2 + x_1^2 = r^2$ and $x_2^2 + x_3^2 = r^2$, corresponding to $\phi = 0$ and $\pi/2$ as shown in figure A1.4. Among these tori, the spherical torus T^2, obtained for $\phi = \pi/4$, is equidistant from the two axes and separates S^3 into two equivalent sub-spaces. Any point on a torus is therefore determined by the variables θ and ω and a constant ϕ. Varying the two first by $d\theta$ and $d\omega$ leads to the orthogonal displacements $r \cos \phi \, d\theta$ and $r \sin \phi \, d\omega$ on the torus and changing the third by $d\phi$ leads to a displacement $r \, d\phi$ normal to the torus. The infinitesimal length dl on the spherical torus reads

$$dl^2 = r^2 (d\theta^2 + d\omega^2)/2$$

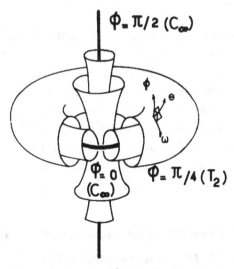

Fig. A1.4. Stereographic projection in R^3 of a family of tori with $\phi = $ constant in S^3. The spherical torus T^2 corresponds to $\phi = \pi/4$, θ and ω are variables on the surfaces of the tori.

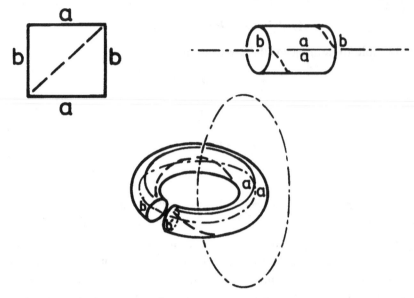

Fig. A1.5. The spherical torus T^2 (represented here by a stereographic projection into R^3) is built upon identifying opposite sides of a square sheet in S^3; as this is done in curved space, the sheet suffers no distortion.

This is a two-dimensional Euclidean metric, which shows that the spherical torus has a zero Gaussian curvature. Indeed it can be built upon identifying the opposite sides of a square sheet of sides $\sqrt{2}\pi r$ as depicted in figure A1.5. Note that, while such a gluing process would require some distortion in R^3, it is done isometrically in S^3.

The element of 2-volume (area) on any torus defined by a constant ϕ (which is also a surface with vanishing Gaussian curvature) reads $da = r^2 \sin\phi \cos\phi \times d\theta \, d\omega$, while the element of 3-volume in S^3 is $dv = r^3 \sin\phi \cos\phi \, d\theta \, d\omega \, d\phi$. The 2-volume of the torus therefore reads $A = 4\pi^2 r^2 \sin\phi \cos\phi$ and its 3-volume (content), $V = 2\pi^2 r^3 \sin^2\phi$. Here again, as in § A1.2.2, the torus separates S^3 into two parts, whose 3-volumes add to $2\pi^2 r^3$. The above formula gives the 3-volume of the part containing the north pole.

Note that the spherical torus T^2 with $\phi = \pi/4$ has an area $2\pi^2 r^2$ and an enclosed volume $\pi^2 r^3$, which is exactly half that of S^3.

A1.3 Two- and three-dimensional cylindrical spaces

A1.3.1 *The standard cylinder $S^1 \times R^1$*

Let us consider first the usual cylinder $S^1 \times R^1$. It can be embedded in a Euclidean three-dimensional space. It is a developable surface with zero

Gaussian curvature and can be considered as an infinite flat strip with two opposite sides identified upon rolling. Geodesic lines on the flat strip are straight lines. By rolling these lines one can get either circles S^1, when the lines are orthogonal to the sides of the strip, or straight lines, when they are parallel to the sides of the strip, and finally helices in the generic case. This space therefore contains both closed and infinite geodesics.

Note that, from an intrinsic point of view, area measurements on the cylindrical surface would indicate a zero Gaussian curvature. The well known system of cylindrical coordinates is the appropriate system to describe this space.

A1.3.2 The three-dimensional cylinder $S^2 \times R^1$

Similar to the case of the above cylinder $S^1 \times R^1$ embedded in R^3, the cylindrical three-dimensional space $S^2 \times R^1$ can be embedded in four-dimensional Euclidean space. It contains several kinds of geodesic: great circles of S^2, straight lines orthogonal to the great circles and also helices. This space has an intrinsic constant positive curvature.

Appropriate cylindrical coordinates are obtained by adding a new parameter to the two-dimensional spherical coordinates. The space can then be described as a foliation of cylinders whose generators are the circles of one family of parallel circles on S^2, including the equator, which leads to a 'great cylinder',

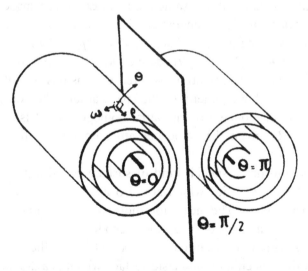

Fig. A1.6. Representation in R^3 of a family of cylinders in $S^2 \times R^1$, defined by $\theta =$ constant. The spherical part (S^2) of the space has been stereographically projected. The 'great cylinder', corresponding to $\theta = \pi/2$, contains the projection pole and is therefore mapped into a plane.

and the two poles, which lead to two straight lines (see figure A1.6). Note that the great cylinder separates the space into two equivalent sub-spaces.

A1.4 Intrinsic curvature

The intrinsic curvature of a space can be determined by direct measurements in the space itself, without any reference or hypothesis about the space of higher dimension into which it is embedded (Cartan 1963). In the case of a two-dimensional surface, this can be done by determining the area of a small disc of geodesic radius ρ. This area is related to the intrinsic curvature, which is also the Gaussian curvature in this case, by $s = \pi\rho^2(1 - \rho^2\kappa/12)$, a formula, valid up to second order, which can be easily tested on a sphere with $\kappa = 1/r^2$. A measure of the length of a small circle can also be used to define the curvature, the appropriate relation reading here: $l = \mathrm{d}s/\mathrm{d}\rho$ or $l = 2\pi\rho \times (1 - \rho^2\kappa/6)$.

Notice that, with this definition, the Gaussian curvature is indeed an intrinsic quantity, as opposed to its more common definition as the product of the two principal curvatures, because in the latter case one needs to embed the surface into a three-dimensional space in order to have the principal curvatures. The two definitions are nevertheless equivalent, provided the three-dimensional embedding space is Euclidean (see Dubois-Violette and Pansu 1990).

In a three-dimensional curved space, the volume content of a small ball differs from that of a ball with the same radius in Euclidean space. This allows us to define a scalar quantity κ characteristic of the intrinsic curvature of the space. Indeed, the volume of a small sphere $V = \frac{4}{3}\pi\rho^3(1 - \kappa\rho^2/15)$, or equivalently its area $A = 4\pi\rho^2(1 - \kappa\rho^2/9)$, defines κ. This definition is clearly an extension of the Gaussian curvature to three dimensions, but it contains no information about the homogeneity of the space around the considered point. In order to clarify this concept of curvature homogeneity, it is worth comparing S^3 and $S^2 \times R^1$, given the same curvature defined by the above scalar quantity κ.

Indeed let us consider a special kind of surface called a geodesic surface. These surfaces have geodesics which are also geodesics of the curved three-dimensional space itself. The need for more precise characteristics of the curvature is clear. In the case of S^3 all these geodesic surfaces through a point are great spheres with a Gaussian curvature $\kappa = 1/r^2$. In the case of $S^2 \times R^1$, through a point there is either a geodesic surface which is a 2-sphere, or others which are cylinders. These cylinders, formed by the product of a great circle of S^2 and R^1, have a vanishing Gaussian curvature.

So it is of great importance to have a tool to characterize this anisotropy. The

Ricci tensor, which is a contraction of the Riemann tensor (Misner *et al.* 1973) has this property. Using appropriate coordinates (spherical for S^3 and cylindrical for $S^2 \times R^1$) the Ricci tensor is a diagonal 3×3 matrix. It is presented in more detail in chapter 4.

In the case of the space S^3 this tensor is:

$$\begin{pmatrix} 2/r^2 & 0 & 0 \\ 0 & 2/r^2 & 0 \\ 0 & 0 & 2/r^2 \end{pmatrix}$$

while in $S^2 \times R^1$ it reads:

$$\begin{pmatrix} 1/r^2 & 0 & 0 \\ 0 & 1/r^2 & 0 \\ 0 & 0 & 0 \end{pmatrix} \tag{A1.6}$$

The trace of the Ricci tensor is the 'scalar curvature' $\mathscr{R}$, the true intrinsic curvature, which is twice the above κ introduced by Cartan: $\mathscr{R} = 6/r^2$ for S^3 and $\mathscr{R} = 2/r^2$ for $S^2 \times R^1$.

A2

Quaternions and related groups

A2.1 Quaternions

A2.1.1 Complex numbers and quaternions

In the eighteenth century, complex numbers were quite often used, but remained mysterious objects. This situation later changed when Gauss, and then Hamilton (around 1835), interpreted these numbers as pairs of real numbers which obey the following addition and multiplication laws:

$$(a, b) + (c, d) = (a + b, c + d)$$
$$(a, b)(c, d) = (ac - bd, ad + bc)$$
(A2.1)

A real number is simply $(a, 0)$ and the imaginary unit becomes $(0, 1)$.

In the following years, Hamilton tried unsuccessfully to generalize this to triplets of real numbers which would satisfy the standard laws of associativity, commutativity and distributivity of multiplication versus addition. One aim was to provide tools for manipulating vectors and rotations in three-dimensional Euclidean space.

In 1843, in what was presented as a 'flash of genius', Hamilton realized that it was possible to build a new consistent system of generalized numbers, represented as quadruplets of real numbers, at the price of losing the commutativity of the multiplication law (du Val 1964).

A2.1.2 'Hamilton' relations

Quaternions are usually presented with the imaginary units $\mathbf{i}$, $\mathbf{j}$, $\mathbf{k}$ in the form:

$$q = q_0 + q_1\mathbf{i} + q_2\mathbf{j} + q_3\mathbf{k}, \quad q_0, q_1, q_2, q_3 \in \mathbf{R}$$
$$\text{with} \quad \mathbf{i}^2 = \mathbf{j}^2 = \mathbf{k}^2 = \mathbf{ijk} = -1$$
(A2.2)

224

the latter 'Hamilton' relations defining the multiplication rules which are non-commutative. Quaternions, which form a corpus **H**, can also be viewed as ordered pairs of complex numbers with the following rules for addition and multiplication:

$$(s, t) + (u, v) = (s + u, t + v),$$

$$(s, t)(u, v) = (su - t\bar{v}, sv + t\bar{u}) \tag{A2.3}$$

where $s, u, v, t \in \mathbf{C}$ and $\bar{u}$ is the complex conjugate of u. The conjugate quaternion is defined as

$$\bar{q} = q_0 - q_1\mathbf{i} - q_2\mathbf{j} - q_3\mathbf{k} \tag{A2.4}$$

and the norm N_q is such that

$$N_q^2 = q\bar{q} = \sum_{i=0}^{3} q_i^2 \geq 0 \tag{A2.5}$$

Another way in which q can be written is as a scalar part $S(q)$ and a vector $\mathbf{V}(q)$:

$$q = S(q) + \mathbf{V}(q), \quad S(q) = q_0, \quad \mathbf{V}(q) = q_1\mathbf{i} + q_2\mathbf{j} + q_3\mathbf{k} \tag{A2.6}$$

with the relations

$$S(q) = \frac{1}{2}(q + \bar{q}), \quad V(q) = \frac{1}{2}(q - \bar{q}) \tag{A2.7}$$

A quaternion is said to be real if $V_q = 0$ and pure imaginary if $S_q = 0$. We shall also write $V_i(q)$ to denote the component of $\mathbf{V}(q)$ along $\mathbf{i}$.

A2.2 Some continuous groups acting on spheres

A2.2.1 The continuous group of unit quaternions

Let Q be the set of unit norm quaternions. It is a non-commutative group isomorphic to $SU(2)$ and to S^3 (considered as a topological group). A quaternion in Q can be written

$$q = \cos \alpha + y \sin \alpha, \quad \text{or} \quad q = \exp \alpha y \tag{A2.8}$$

where y is a pure imaginary quaternion. Recall that quaternion multiplication is not commutative so that

$$\exp \alpha y \exp \beta z = \exp(\alpha y + \beta z) \tag{A2.9}$$

only if $y = z$.

Each element of Q is in one to one correspondence with a point on the unit radius S^3 sphere. The space S^3 is a 'topological group', e.g. each point of that space represents an element of the group. Because Q is a group, it can also label displacements on S^3. This is very similar to the relation between the group of unit complex numbers (isomorphic to $SO(2)$, the group of planar rotations) and the unit circle (S^1) in the plane. But in the quaternion case, there is the additional complexity of the non-Abelian nature of Q.

A2.2.2 Displacement on a sphere

Let us describe the metric preserving symmetry properties of the sphere S^2 in R^3. Very general transformations in R^3 are provided by the elements of the linear group $GL(3, R)$ which are 3×3 matrices with real elements as indicated by the arguments of GL. The sphere S^2 is allowed to be globally invariant by the elements of the orthogonal group $O(3)$, sub-group of $GL(3, R)$, consisting of 3×3 matrices with determinant equal to ± 1 and such that the inverse of a matrix is equal to its transpose. Those transformations which preserve orientation on S^2 are elements of $SO(3)$, a sub-group of $O(3)$ containing the matrices with determinant equal to $+1$.

A rotation in three dimensions can be expressed easily in terms of quaternions. Consider the rotation of angle 2α around an axis given by a vector **y**. Build the unit pure quaternion

$$y = y_1\mathbf{i} + y_2\mathbf{j} + y_3\mathbf{k}$$

The rotation is then given by the transformation

$$x \rightarrow e^{-\alpha y}xe^{\alpha y} \tag{A2.10}$$

Note that a given rotation in R^3 corresponds to two quaternions (in the above example, $e^{\alpha y}$ and $e^{(\alpha+\pi)y} = -e^{\alpha y}$), which shows the 2:1 homomorphism between the unit quaternions and $SO(3)$.

It was not before the end of the nineteenth century that quaternions were considered as points (and symmetries) in four dimensions: 'they play the role of four dimensional eyes'. The group $SO(4)$ represents transformations on the sphere S^3. Let us now characterize displacements on S^3. The quaternion q will represent a point on S^3, while the unit quaternions l, r are the displacements. Owing to the non-commutativity, one must distinguish quaternions acting on the left (l) and on the right (r). The transformation qr is a direct isometry, called a right screw. If r is written in the form $r = e^{\alpha y}$, and if α is varied from 0 to 2π, the orbit of q describes a great circle of S^3. One can similarly define left screws. These screws can also be seen as simultaneous rotations of the same angle α around two completely orthogonal great circles. Here, a rotation is an operation which leaves invariant a two-plane in R^4. Under a given screw, the orbits of two points either coincide or are completely disjoint, forming a great circle fibration of S^3 (a Hopf fibration, see appendix A3). As screws can be seen as combinations of rotations, rotations can also be defined as combinations of screws. A rotation of angle α can always be written as lqr with $S(l) = S(r) = \cos(\alpha/2)$, their vectorial part defining the invariant plane. In addition, one can show that any element of $SO(4)$, acting on q, can be represented by lqr with $S(l) \neq S(r)$ in the generic case. Finally, the indirect orthogonal transformations can also be defined, like the 'rotation–reflection', $q \rightarrow -\bar{l}\bar{q}r$ with $\pm l \neq r$.

The different continuous symmetry groups (of interest in the present context) display the following inter-relations :

$$S^3 = Q = SU(2)$$

$$SO(3) = SU(2)/Z_2 \tag{A2.11}$$

$$SO(4) = SU(2) \times SU(2)/Z_2$$

where Z_2 is the two-element group and $SU(2)$ the group of unitary unimodular 2×2 matrices.

A2.3 Discrete groups

In Euclidean three-dimensional space, the polyhedral symmetry groups are finite sub-groups of $SO(3)$, the continuous group of direct rotations. However $SO(3)$ is not simply connected and is double-covered by the group Q as indicated in §A2.2.2. This 2:1 homomorphism is extended to the finite groups, the pre-image of a standard polyhedral group being called a binary polyhedral group (or a double group), which can then be represented as a discrete group of unit quaternions.

A2.3.1 The Hurwitz group

Let us first describe the so-called Hurwitz group of unit quaternions T, also called the binary tetrahedral group. It contains the following 24 elements

$$T = \{\pm 1, \pm \mathbf{i}, \pm \mathbf{j}, \pm \mathbf{k}, \tfrac{1}{2}(\pm 1 + \pm \mathbf{i} + \pm \mathbf{j} + \pm \mathbf{k})\} \tag{A2.12}$$

If one considers the group elements as points on the unit sphere S^3, they form a

{3, 4, 3} self-dual polytope. Note also that the set of vertices of the {3, 3, 4, 3} tessellation, taken as vectors, form a ring whose units are precisely the elements of T (du Val 1964).

A2.3.2 The binary icosahedral group

The next very interesting group is the binary icosahedral group Y' (the prime denotes the double group case) which is related to the polytope {3, 3, 5} as explained below (see also Warner 1982). The image in Q of the dihedral group D_2 (of order 4) is the group V' (of order 8) consisting of the eight principal unit quaternions:

$$(\pm1, 0, 0, 0), \quad (0, \pm1, 0, 0), \quad (0, 0, \pm1, 0), \quad (0, 0, 0, \pm1) \qquad \text{(A2.13)}$$

The above presented binary tetrahedral group (the Hurwitz group) can be constructed from the group V':

$$T' = \oplus^2_{r=0} \left(\frac{1}{2}, \frac{1}{2}, \frac{1}{2}, \frac{1}{2}\right)^r . V' \qquad \text{(A2.14)}$$

where $\oplus^2_{r=0} A_r$ means the union of n sets A_r which have no points in common and with the convention that a quaternion raised to the power zero equals $(1, 0, 0, 0)$, the neutral element of Q.

The group Y' can then be written

$$Y' = \oplus^4_{s=0} T' . \left(\frac{1}{2}\tau^{-1}, \frac{-1}{2}\tau, 0, \frac{-1}{2}\right)^s \qquad \text{(A2.15)}$$

Here τ is the golden ratio. The 120 elements of Y' correspond to the 120 vertices of the {3, 3, 5} polytope on a unit radius hypersphere S^3 (see appendix A5).

The direct symmetry group of the {3, 3, 5} polytope is a sub-group of $SO(4)$ which reads

$$G' = Y' \times Y'/Z_2 \qquad \text{(A2.16)}$$

The two-element group Z_2, with $(\pm1, 0, 0, 0)$ as elements, commutes with all elements of Q. The quotient by Z_2 takes this into account, identifying for example the transformations $-lqr$ and $lq(-r)$.

Since the order of Y' is 120, the quotient by Z_2 implies that the order of G' is 7200. The total symmetry group G also includes indirect orthogonal transformations, analogous to reflections for ordinary three-dimensional discrete groups. These are given by

$$q = \bar{l}\bar{q}r^{-1}l, \quad r \in Q$$

This adds 7200 new elements and gives the full group G of order 14 400.

A2.3.3 The polytope '240' symmetry group

This polytope is described in chapter 2 as a template for tetravalent amorphous materials. It is obtained from two replicas of the $\{3, 3, 5\}$ polytope related by a screw. Its direct symmetry group is

$$G_{240} = Y' \times O'/Z_2 \tag{A2.17}$$

where O' is the lift of the octahedral group. Note the two following points. Firstly, O' is not a sub-group of Y'; and polytope '240', while sharing some of the $\{3, 3, 5\}$ symmetries, also has new symmetries, in particular a 40-fold screw axis. Secondly, polytope '240' is chiral: it cannot be superimposed onto its mirror image. The polytope '240' with opposite chirality has $G = O' \times Y'/Z_2$ as its symmetry group.

A3

Hopf fibration

A3.1 Fibrations

A space can be considered as a fibre bundle if there is a sub-space (the fibre) which can be reproduced by a displacement so that any point of the space is on a fibre, and only one. For example the Euclidean space R^3 can be considered as a fibre bundle of parallel straight lines, all perpendicular to the same plane.

A3.1.1 What is a fibration?

From a mathematical point of view, a fibred space E is defined by a mapping p from E onto the so-called 'base' B, any point of a given fibre being mapped onto the same base point. A fibre is therefore the full pre-image of one base point under the mapping p. In a three-dimensional fibred space with one-dimensional fibres, the base is a two-dimensional manifold. In the above simple R^3 example, the two-dimensional base space is just the plane orthogonal to the fibres; in this case the base is a sub-space of the whole space, and the fibration is called 'trivial'. But this latter property is not general, the base may not be embedded in the fibre bundle space, as will be seen below with the Hopf fibration.

A3.1.2 Fibration of $S^2 \times R^1$

When a space is the direct product between two spaces, it can trivially be considered as a fibre bundle. If fibres are straight lines in $S^2 \times R^1$ then the base is the sphere S^2. In the simple case of $S^1 \times R^1$ (the classical cylinder), fibres are generatrices, but it is also possible to consider helicoidal fibres; in this case the base S^1 is not orthogonal to the fibres (figure A3.1).

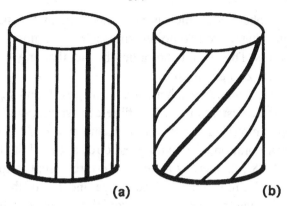

(a)　　　　　　　　　**(b)**

Fig. A3.1. Fibration of $S^1 \times R^1$ by straight lines (a) and helices (b). This is an example of a trivial fibred space.

A3.2 Hopf fibration

There are two well known Hopf fibrations of (high-dimensional) spheres by (lower-dimensional) spheres. The most famous is the fibration of S^3 by great circles S^1 and base S^2. It is a nice example of a non-trivial fibration ($S^3 \neq S^2 \times S^1$). This fibration can be extended to the whole R^4 as a bundle of 2-planes which only meet at the origin. It is used several times in this book, as in the description of the unfrustrated blue phase (chapter 2) or in order to locate disclination lines in polytopes (chapter 4). The second fibration, that of S^7 with fibres S^3 and base S^4, is used here in a discretized form for a description of the $E8$ lattice (Sadoc and Mosseri 1993).

A3.2.1 Fibration of S^3

Let us write the points on a unit radius S^3 as pairs of complex numbers (u, v) such that $u^2 + v^2 = 1$. The Hopf map may then be defined as the composition of the map h_1 from S^3 to R^2 (with ∞ included) followed by an inverse stereographic projection from R^2 to S^2:

$$h_1: \quad S^3 \rightarrow R^2$$

$$(u, v) \rightarrow Q = uv^{-1} \quad u, v \in \mathbf{C}$$

$$h_2: \quad R^2 \rightarrow S^2$$

$$Q \rightarrow M = (x_1, x_2, x_3) \quad x_i \in \mathbf{R}$$

(A3.1)

The base is a 2-sphere S^2, which is not embedded in S^3. Indeed, if it was embedded, it would then have two points intersected by a fibre, as a circle cuts a sphere in two points. This would contradict the fact that only one point on the base should characterize a given fibre.

Let a base point be defined by spherical coordinates θ_o and ϕ_o and let us use the toroidal coordinate system on S^3. The fibre corresponding to the base point is a great circle parametrized by: $\theta = \omega + \theta_o$ and $\phi = \phi_o/2$. The full inverse image of S^2 gives a fibration of S^3 with great circles, called the Hopf fibration. A circle on the base S^2, defined by a constant ϕ_o, represents a torus in S^3. The sequence of 'parallel' circles on S^2 is the image under p of the torus foliation of S^3, which appears naturally in the S^3 toroidal coordinate system: a torus is defined by a constant ϕ parameter. This torus foliation was previously mentioned in appendix A1. Note that this foliation has two particular members: $\phi = 0$ and $\pi/2$ correspond to two great circles, members of the Hopf fibration, which are the axes of the torus foliation. The case $\phi = \pi/4$ corresponds to the 'spherical torus'. See figures A3.2 and A3.3.

Each of these tori can be seen as a two-dimensional space equivalent to a

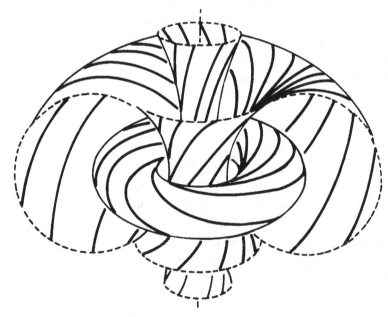

Fig. A3.2. Hopf fibration of S^3, by stereographic mapping onto R^3. The fibres are great circles, some of which are drawn here on selected tori of the S^3 torus foliation. The whole family of these great circles of S^3 is also called a family of Clifford parallel lines.

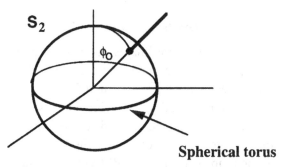

Spherical torus

Fig. A3.3. The base S^2 of the Hopf fibration. A point is the image of a fibre. The equator corresponds to the spherical torus. The following relations apply: $\psi: S^3 \rightarrow S^2$ or with coordinates: $(\theta, \phi, \omega) \rightarrow (\alpha, \eta)$ with $\alpha = 2\phi$, $\eta = \theta - \omega[2\pi]$, $\alpha \in [0, \pi]$, $\eta \in [0, 2\pi[$.

rectangle with opposite sides identified two by two (or a square in the particular case of the spherical torus). All diagonals of these rectangles have the same length and correspond to S^3 great circles (upon identification of the opposite sides which close these lines). So, the set of lines parallel to a rectangle diagonal corresponds to a family of great circles, which do not intersect in S^3. All these great circles drawn on the torus foliation are a Hopf fibration of S^3 (figure A3.4).

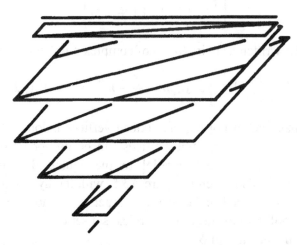

Fig. A3.4. If the torus is unfolded into rectangles, the fibration appears as a family of segments parallel to diagonals of rectangles. This gives an idea of the torsion of the Hopf fibration. All diagonals have the same length, and they are refolded into great circles of S^3 when the rectangles are refolded into tori.

A3.2.2 The Hopf map from S^7

A point on the sphere S^7 has its coordinates given by a pair of quaternions (q_1, q_2) subject to $q_1 \bar{q}_1 + q_2 \bar{q}_2 = 1$. S^7 can be fibred such that the quaternion $Q = q_1 q_2^{-1}$ is invariant on each fibre S^3, and therefore uniquely specifies the fibre. Q can be any quaternion, including ∞, which shows that the base space of the bundle is S^4. The latter can be viewed as the inverse stereographic projection from R^4 to S^4. In terms of coordinates, the Hopf map can be written as the composition of the map h_1 from S^7 to R^4 (with ∞ included) followed by an inverse stereographic projection from R^4 to S^4. There are strong similarities with the S^3 fibration, but here quaternions replace complex numbers. The interested reader can find more details in two references (Manton 1987, Sadoc and Mosseri 1993).

A3.2.3 Discretized Hopf fibrations

Let us now come to the discretized Hopf fibrations. They have already been used in the context of curved space models for the description of polymer packing (Kléman 1985a, b) and, in a quite different field, in order to introduce disclination lines into polytopes (Nicolis *et al.* 1986).

Let us first look at the simple example of a hypercube in R^4. Its sixteen vertices read

$$\left\{ \frac{1}{2} (\pm 1, \pm 1, \pm 1 + \pm 1) \right\}$$

which can also be written as pairs (u, v) of complex numbers

$$\left\{ \left(\frac{1}{\sqrt{(2)}} e^{i\theta}, \frac{1}{\sqrt{(2)}} e^{i\omega} \right) \right\}, \quad \theta, \omega = \frac{\pi}{4} + n\frac{\pi}{2} \quad n = 0, 1, 2, 3 \qquad \text{(A3.2)}$$

Since u and v have equal norms, the sixteen vertices belong to the spherical torus. On this torus, the hypercube vertices can be grouped along four lines (images of four Hopf circles) parallel to the square diagonal. A three-dimensional image is also given in figure A3.5, with the hypercube depicted in the familiar form of a small cube inside a larger one (known as the Schlegel diagram). The Hopf map of these four circles is a set of four points symmetrically located on the equator of S^2.

Consider now the $\{3, 3, 5\}$ polytope. Since each screw symmetry gathers vertices on Hopf circles, different discretized fibrations can be drawn on this polytope, which we now describe.

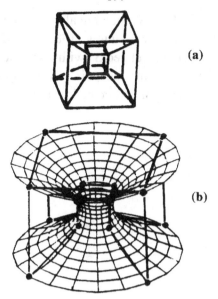

Fig. A3.5. A three-dimensional representation of a hypercube (a), which appears as a small cube inside a larger one (this is called a Schlegel diagram). (b) The hypercube vertices can be grouped on a torus, on which they form a square tiling. Vertices can then be gathered onto Hopf circles parallel to a square diagonal.

- The ten-fold screw axis: there are 12 fibres (containing ten points each), whose Hopf map gives the 12 vertices of an icosahedron on the base S^2.
- The six-fold screw axis: there are 20 fibres (containing six points each), whose Hopf map gives the 20 vertices of a dodecahedron on S^2.
- And finally the four-fold screw axis with 30 fibres (containing four points each), whose Hopf map gives the 30 vertices of an icosidodecahedron on S^2.

Note that the order of the screw symmetry (ten, six and four), which is the order of the regular polygon on each circular fibre, is exactly twice the order of rotation symmetry around a vertex of the polyhedron in the Hopf map. This latter symmetry corresponds in fact to the local symmetry among the fibres in S^3. If a point on S^2 has five neighbours, it means that the corresponding fibre in S^3 has five other fibres nearby. It is this property which has been exploited in order to get disclinated polytopes, by first disclinating the base polyhedron (see chapter 4).

Analogously, if one is interested only in the geometry of a discrete set on S^7 (a polytope), one can restrict oneself to a finite set of S^3 fibres, each containing a discrete set of points. When the set on S^7 is highly symmetrical, it is possible to choose the orientation of the discrete bundle such that each of the S^3 fibres contains a symmetrical set, the base S^4 showing also a symmetrical pattern. For example, as explicitly described by Manton (1987), the 240 vertex Gosset polytope on S^7 is fibred into ten discretized S^3 with 24 vertices each, forming a $\{3, 4, 3\}$ polytope. The base contains ten points forming a 'cross' polytope (the higher-dimensional equivalent of an octahedron) on S^4 (see appendixes A4 and A9).

A4

Polytopes and honeycombs

A4.1 Symmetries and orthoscheme tetrahedra

A4.1.1 Group generated by reflections. The characteristic triangle

Let us consider a triangle ABC, with angles α, β, γ, located on any surface, either Euclidean, spherical or hyperbolic, and let a, b, c denote the simple reflections in the sides BC, AC and AB (du Val 1964). In the following we shall embed the spherical geometry in R^3 as the standard sphere S^2. ABC becomes in this case a spherical triangle, whose sides are great circle arcs; a, b and c are then the simple reflections in the planes defined by these great circles. In the hyperbolic space case, we shall use the unit disc conformal model (appendix A1). The triangle sides are now arcs of circles orthogonal to the absolute (these lines can degenerate into straight parts of a diameter when the geodesics run through the origin), and reflections in the sides are replaced by inversion with respect to these circles (figure A4.1).

Let us first briefly recall what a circle inversion is. Consider a circle of centre O and radius r in the Euclidean plane and a point M anywhere in that plane. The inverse M' of M with respect to the circle is located on the line OM at a position such that $OM.OM' = r^2$.

The three types of triangle are represented in figure A4.2.

It is possible to generate rotations by combining the reflections. For example the symmetry operations obtained as the products $b.c$, $c.a$ and $a.b$ are rotations of angles 2α, 2β and 2γ about A, B and C respectively. Now assume that $\alpha = \pi/n$, $\beta = \pi/p$ and $\gamma = \pi/q$ with n, p and q integers. From the classical relation for the sum of the angles of a triangle, according to its hyperbolic, Euclidean or spherical nature, we have:

$$1/n + 1/p + 1/q < 1, \quad = 1 \text{ or } > 1 \qquad (A4.1)$$

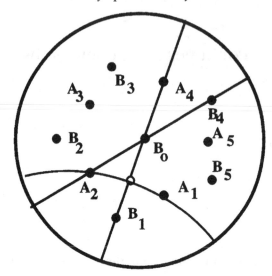

Fig. A4.1. The characteristic triangle associated with the pentagon of the $\{5,4\}$ hyperbolic tessellation. The A_i are the vertices of the central pentagon. They are transformed into each other by inversion through the sides of the triangle, as for the B_i which are the centres of the pentagons (or the vertices of the dual tiling!).

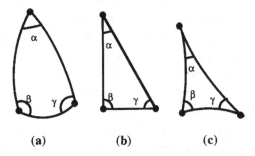

$$\text{(a)} \qquad\qquad \text{(b)} \qquad\qquad \text{(c)}$$

Fig. A4.2. Characteristic triangles in the three examples of two-dimensional geometries: (a) $\alpha + \beta + \gamma > \pi$, spherical space with positive curvature; (b) $\alpha + \beta + \gamma = \pi$, Euclidean flat space; (c) $\alpha + \beta + \gamma < \pi$, hyperbolic space with negative curvature.

The last two cases offer only a limited number of possibilities. The only solutions for the Euclidean plane are:

$$(n, p, q) = (3, 3, 3), (2, 4, 4), (2, 3, 6)$$

On S^2 we get:

$$(n, p, q) = (n, 2, 2)n > 2, (2, 3, 3), (2, 3, 4), (2, 3, 5) \qquad (\text{A4.2})$$

For all the other sets of values (n, p, q), the pattern can only exist in the hyperbolic plane. The reflections a, b, c generate a group, which includes reflections in all the edges of the triangular tiling, as in the mirrors of a kaleidoscope (Coxeter 1973a, b, du Val 1964). This group, which is a discrete sub-group of all motions and reflections in the two-dimensional manifold, is simply transitive on the triangles and is called the triangle group $(n, p, q)^*$. It has a pure sub-group (n, p, q) containing orientation preserving transformations. Correspondingly, it is possible to divide the triangles into two sets, according to whether they are obtained from ABC by an odd or an even number of reflections. The pure group (n, p, q) is simply transitive on each set of triangles separately. The groups generated by reflections have been extensively discussed by Coxeter and Moser (1957). Regular sets can now be defined by giving the group which let them be invariant and the location of vertices in one of the triangles which will be called the fundamental triangle. This is similar to standard crystallography where a crystal can be characterized by the location of atoms in the unit cell, and by the translation part of the full crystal group. Note that usually crystallographers take advantage of point group symmetry and then characterize the unit cell by specifying a smaller number of atomic positions. In the present language, it consists of dividing the unit cell into fundamental tetrahedra, the full group of the regular set being generated by reflections in the faces of these tetrahedra.

In the case of spherical tessellations, $(n, p, q)^*$ and (n, p, q) are respectively sub-groups of $O(3)$ and $SO(3)$. Then $(2, 3, 3)^*$, $(2, 3, 4)^*$ and $(2, 3, 5)^*$ are the full tetrahedral, octahedral and icosahedral groups. The patterns of triangles on S^2 are shown in figure A4.3. Note that the number of triangles is equal to the order of the full group, which is here respectively 24, 48 and 120 and twice the order of the pure sub-groups.

Figure A4.4 shows two examples of hyperbolic tessellations associated with the groups $(2, 3, 7)^*$ and $(2, 4, 6)^*$. These kinds of tessellations have inspired some drawings of the Dutch artist Escher.

We are interested, throughout this book, in approximating a two-dimensional manifold by a collection of flat faces. This is called a polyhedral approximation or a faceting. Let us consider the three groups

$$(2, 3, 4)^*, (2, 4, 4)^*, (2, 5, 4)^*$$

which act respectively on S^2, R^2 and H^2. If the fundamental triangle contains only one site located at B, the orbit under the group gives rise to the square tessellations $\{4, 3\}$, $\{4, 4\}$ and $\{4, 5\}$.

Assume now that these three tessellations have the same flat basic squares (the reader is urged to take some glue, a set of similar paper squares and do the

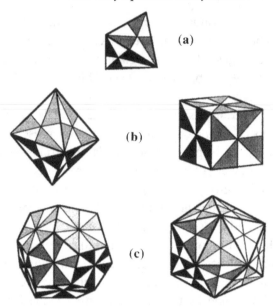

Fig. A4.3. Patterns of triangles corresponding to the polyhedral sub-groups of $O(3)$. (a) $(2, 3, 3)^*$; (b) $(2, 3, 4)^*$; (c) $(2, 3, 5)^*$. The triangle pattern is shown on S^2 and mapped onto the faces of the regular polyhedra which share the same symmetry group. In each case the pure sub-group (n, p, q) is transitive on the triangle pattern with fixed colour.

construction!). They differ now by the number of such squares shared by any vertex. Gluing three squares at a vertex, one eventually gets a cube, a poly-hedral approximation of S^2. Four squares around each vertex naturally gives the square lattice of R^2. The most interesting exercise consists of trying to attach five squares around each vertex. This is locally possible if one escapes in the third dimension (figure A4.5a), in which case it is possible to propagate this local order for a while. The local rule 'put five squares around each vertex' can be obeyed indefinitely while constructing the so-called three-dimensional polyhedra (Wells 1977) but the underlying space is no more simply connected: there are closed circuits which do not separate the space into two disjoint pieces. Figure A4.5b displays locally one such infinite polyhedron where the above mentioned circuits enclose the (square base) tubular regions. This means that even if the hyperbolic $\{4, 5\}$ and this infinite three-dimensional polyhedron share the same local rule, they differ from a topological point of view.

In the polyhedral models, the curvature is concentrated at the vertices with cone-like (positive curvature) or saddle-like (negative curvature) shapes. It is

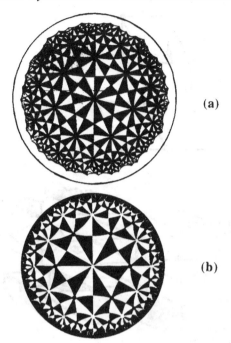

(a)

(b)

Fig. A4.4. Patterns of triangles corresponding to the hyperbolic groups. (a) (2, 3, 7)[*] (after Magnus 1974); (b) (2, 4, 6)[*] (after Coxeter 1969). The unit disc model of the hyperbolic plane is used.

encoded in the so-called 'deficit angle' at the vertex which measures the deviation from 2π of the sum of the internal angles of the polygons meeting at the vertex. It is $\pi/2$ for the cube $\{4, 3\}$ and $-\pi/2$ for the hyperbolic tessellation $\{4, 5\}$. This is nothing other than the Gauss–Bonnet relation presented in chapter 4. It is worth noticing the following sum rule for the angular deficit δ_i at the ith vertex of a polyhedron with a two-dimensional spherical topology:

$$\sum \delta_i = 4\pi$$

In higher dimension it is possible to generalize the polyhedral approximation, as has been shown by Regge (1961) in his attempt to provide a discrete description of curved geometries in general relativity (see chapter 4). For example, a three-dimensional curved space can be approximated by a simplicial (tetrahedral) packing, the curvature being concentrated along the edges. But, owing to the absence of a Gauss–Bonnet-like relation in higher-dimensional space, these discretized spaces do not faithfully reproduce the continuous geometries.

(a)

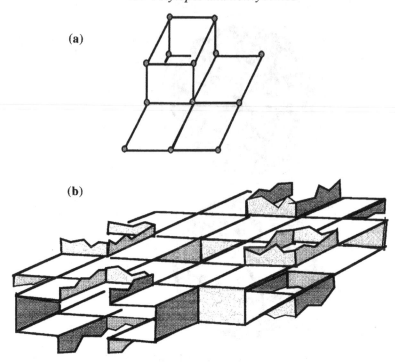

(b)

Fig. A4.5. Tiling with five squares around a vertex. (a) Approximation of the hyperbolic tiling {4,5} by gluing squares in R^3. (b) An infinite three-dimensional polyhedron such that every vertex belongs to five squares. The topology is not globally equivalent to the hyperbolic case (see Wells 1977).

A4.1.2 Order of the two-dimensional polyhedral groups

Let us consider the characteristic, or Möbius, triangle on S^2. All identical triangles obtained by reflection in their sides form a pattern covering the sphere just once.

The order h of a triangle group $(2, p, q)$ is the number of such Möbius triangles that cover the sphere. For a unit sphere of area 4π, using the area of the spherical Möbius triangle it becomes:

$$h = \frac{4}{\dfrac{1}{p}+\dfrac{1}{q}-\dfrac{1}{2}} = \frac{8pq}{4-(p-2)(q-2)} \qquad \text{(A4.3)}$$

It is quickly verified that the full icosahedral group $(2, 3, 5)^*$ has order 120 as previously indicated. The icosahedron, or the dodecahedron, each have less than 120 vertices; this comes from the fact that both are the image, under the action of the group elements, of a special point on the characteristic triangle.

A4.1.3 The orthoscheme tetrahedron

The orthoscheme tetrahedron is a very efficient way to describe a polytope on S^3. It is a generalization in higher dimension of the above characteristic triangle. The faces of this tetrahedron act as mirrors which are combined in order to generate the element of the polytope symmetry group. Faces of the orthoscheme tetrahedron are rectangular triangles defined on the geodesic surface of S^3. As for the two-dimensional case, this figure acts as a kaleidoscope in order to generate polytopes.

Note that orthoscheme tetrahedra are not restricted to describing finite polytopes, but can also be defined for infinite (Euclidean or hyperbolic) groups. As an example, the orthoscheme tetrahedron for the $\{4, 3, 4\}$ is represented in figure A4.6. It is the simplest three-dimensional tessellation: the Euclidean cubic tessellation. All the symmetries of the cubic lattice are generated by the orthoscheme four mirror faces. Dihedral angles of the orthoscheme are related to the rotational symmetries, since the product of two reflections is a rotation. There are three dihedral angles which are π/p, π/q, π/r for a $\{p, q, r\}$ polytope (see figure A4.7, where angles are depicted, following Coxeter notations).

Φ_{pq}, Ψ_{pq}, χ_{pq} (respectively Φ_{qr}, Ψ_{qr}, χ_{qr}) are the sides of the $\{p, q\}$ (or $\{q, r\}$, respectively) cell. They are obtained from the relations:

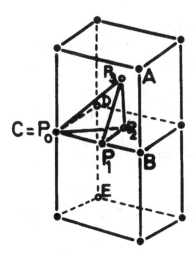

Fig. A4.6. The orthoscheme tetrahedron of the Euclidean $\{4, 3, 4\}$ tessellation. The four vertices of this tetrahedron are respectively the centre of the polytope cell (here the cube), a vertex of this cell, a mid-edge point and finally a (square) face centre. They are labelled P_0, P_1, P_2, P_3 according to the dimension of the n-face (in generic sense) which they are centring.

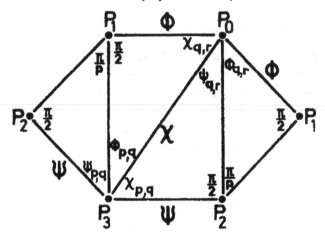

Fig. A4.7. The orthoscheme tetrahedron is here represented unfolded in order to display all its important angles.

$$\cos \chi_{pq} = \cot \frac{\pi}{p} \cot \frac{\pi}{q}$$

$$\sin \Phi_{pq} = \sin \chi_{pq} \sin \frac{\pi}{p} \tag{A4.4}$$

$$\sin \Psi_{pq} = \tan \Phi_{pq} \cot \frac{\pi}{p}$$

and similarly with q and r indices.

Φ, Ψ, χ are the lengths of the orthoscheme tetrahedron edges which are related to the dihedral angles π/p, π/q, π/r:

$$\cos \Phi = \cos \frac{\pi}{p} \sin \frac{\pi}{r} \Big/ \sin \frac{\pi}{h_{qr}}$$

$$\cos \Psi = \sin \frac{\pi}{p} \cos \frac{\pi}{r} \Big/ \sin \frac{\pi}{h_{pq}} \tag{A4.5}$$

$$\cos \chi = \cos \frac{\pi}{p} \cos \frac{\pi}{p} \cos \frac{\pi}{r} \Big/ \sin \frac{\pi}{h_{qr}} \Big/ \sin \frac{\pi}{h_{pq}}$$

where h_{pq} is given by

$$\cos^2 \frac{\pi}{h_{pq}} = \cos^2 \frac{\pi}{p} + \cos^2 \frac{\pi}{q} \tag{A4.6}$$

With these relations, all elements of the orthoscheme tetrahedron are determined in terms of p, q, r. Extension to the hyperbolic case is achieved by mainly replacing trigonometric functions by hyperbolic ones.

A4.2 Polytopes and honeycombs

A4.2.1 Regular structures in curved space

Regular tilings, which are associated with the above groups, can be defined. They are called polytopes in the finite cases and honeycombs otherwise. A three-dimensional curved space with constant curvature can be tiled with regular polyhedral cells. Here the word 'regular' means that all the faces of the polyhedron are regular and equal, each vertex sharing the same number of faces. The standard notation, which covers the various cases, is the Schläfli symbol $\{p, q, r\}$. The $\{p, q\}$ denotes the cell, i.e. a regular polyhedron: p is the symmetry order of a face ($p = 3$ for an equilateral triangle, $p = 4$ for a square), q the number of faces around a vertex in a cell ($p = 3$, $q = 3$ denotes a regular tetrahedron, $\{4, 3\}$ a cube). The duality between polyhedra is clear from the notation: p, q and q, p are dual polyhedra. Then, $\{p, q, r\}$ describes an assembly of $\{p, q\}$ polyhedral cells, r of which share a common edge. The coordination polyhedron around a vertex is the vertex figure defined by the two last indices $\{q, r\}$.

There are five regular polyhedra in R^3,

$$\{3, 3\} \quad \{4, 3\} \quad \{3, 4\} \quad \{3, 5\} \quad \{5, 3\}$$

the tetrahedron, cube, octahedron, icosahedron and dodecahedron, respectively. Plato used the first four such solids to represent the four natural elements (air, fire, earth and water) and the fifth (the only one that contains pentagons) to symbolize the universe. The fact that there are only five regular polyhedra can be seen as follows. The characteristic angle of a regular p-gon is $(1 - 2/p)\pi$; since q of them have to fit around a common vertex, and since the angle of a spherical $\{p\}$ is greater than $(1 - 2/p)\pi$, this yields the inequality $(p - 2)(q - 2) < 4$, which is only satisfied in the five above cases (note that we have imposed that p and q be greater than 2). All other sets $\{p, q\}$, which do not satisfy this inequality, either describe a regular honeycomb of the Euclidean plane

$$(p - 2)(q - 2) = 4 \rightarrow \{4, 4\} \quad \{3, 6\} \quad \{6, 3\} \tag{A4.7}$$

which gives the square, triangular and hexagonal tilings, or a regular honeycomb of the hyperbolic (Lobatchevskyan) plane: $(p - 2)(q - 2) > 4$, which gives infinitely many tilings. In other words, the Schläfli symbols $\{p, q\}$

describe all the regular tilings of homogeneous two-dimensional spaces, which take either the form of polyhedra or of infinite honeycombs.

Now, let us put r regular polyhedra around a common edge. The possible tilings are described by Schläfli's trigonometrical condition (Coxeter 1973a, b), which reads as follows. The quantity

$$\sin\frac{\pi}{p}\sin\frac{\pi}{r} - \cos\frac{\pi}{q} \tag{A4.8}$$

is positive for a spherical space tiling, negative for a hyperbolic space tiling and null for a Euclidean space tiling. One finds:

– six spherical polytopes

$$\{3, 3, 3\} \ \{3, 3, 4\} \ \{4, 3, 3\} \ \{3, 4, 3\} \ \{3, 3, 5\} \ \{5, 3, 3\}$$

– one Euclidean honeycomb, made of cubes

$$\{4, 3, 4\}$$

– eight hyperbolic honeycombs

$$\{6, 3, 3\} \ \{5, 3, 4\} \ \{6, 3, 4\} \ \{4, 4, 3\} \ \{4, 3, 5\} \ \{3, 5, 3\} \ \{5, 3, 5\} \ \{6, 3, 5\}$$

Notice that some hyperbolic honeycombs do not have their dual honeycomb in this list, which is reduced to those having a fundamental domain of finite size, except a limit case for the $\{6, 3, 3\}$. Let us describe briefly this honeycomb and its dual $\{3, 3, 6\}$. The $\{6, 3, 3\}$ is a tessellation by 'polyhedra' ($\{6, 3\}$) which are a limit case since they correspond to hexagonal plane tilings. Here, it is a hexagonal tiling of a so-called 'horosphere', a strange geometrical object existing in hyperbolic space but whose metric is Euclidean. So, it is a rather peculiar honeycomb. But its dual is even more so: its vertex figure $\{3, 6\}$ is infinite, which means that its vertices have an infinite coordination number and lie at infinity! This is why we kept it out of the above list.

Among the six regular polytopes, the first three should be distinguished in the sense that they have trivial generalizations in higher dimensions: $\{3, 3, 3\}$ is the simplex, $\{4, 3, 3\}$ is the hypercube and $\{3, 3, 4\}$ the cross polytope (the generalization in higher dimension of the octahedron). They are often called α_n, γ_n and β_n in n dimensions.

In addition to regular honeycombs or polytopes, it is sometimes interesting to consider semi-regular polytopes or honeycombs. They share the symmetry group of the regular structures but correspond to the image under the group action of more generic points in the orthoscheme.

A4.2.2 The Coxeter statistical honeycomb

There is no Euclidean tiling made of tetrahedra $\{3, 3\}$, since the $\{3, 3, 5\}$ is spherical and the $\{3, 3, 6\}$ is hyperbolic (and complicated as discussed above!). Coxeter proposed a particular 'statistical' honeycomb with a non-integer number of tetrahedra sharing an edge. This structure is quite interesting in the present context since it represents the simplest approximation of a Euclidean tetrahedrally close packed structure.

Schläfli's trigonometrical condition (eq. A4.8) can be forced to zero with $p = q = 3$, giving $r = 5.1043$. This leads to a coordination number $z = 13.397$, using $z(6 - r) = 12$. This value is very close to what is found in real Frank–Kasper crystals (chapter 7), or even as the number of faces in soap froths (Kusner 1992). These numerical values clearly result from a crude approximation. This is in contrast with the two-dimensional case where exact average values can be calculated. In any infinite triangular tiling (regular or not), one exactly finds an average of six triangles meeting at a vertex. In the three-dimensional case, around the above approximate values ($r \simeq 5.1043$ and $z \simeq 13.397$), there are definite fluctuations which depend on the details of the structure (see appendix A6 and §4.3.3).

A5

Polytope $\{3, 3, 5\}$

A5.1 The geometry of polytope $\{3, 3, 5\}$

A5.1.1 Polytopes and the Schläfli notation

A general presentation of polytopes and honeycombs has been given in appendix A4. Here, we aim to describe in some details the $\{3, 3, 5\}$ regular polytope. If the edge length is taken as unity, the polytope is inscribed on a hypersphere of radius equal to the golden ratio, $\tau = (1 + \sqrt{5})/2$. Let us recall the meaning of the Schläfli notation in the case of the $\{3, 3, 5\}$ polytope: it denotes a regular structure such that five $\{3, 3\}$ tetrahedra share a common edge. The first neighbour configuration around a vertex is an icosahedron coded in the two last numbers $\{3, 5\}$.

The $\{3, 3, 5\}$ polytope contains altogether: 120 vertices, 720 edges, 1200 triangular faces, and 600 tetrahedral cells.

Note in passing that the generalized Euler–Poincaré relation

$$V - E + F - C = 0 \qquad (A5.1)$$

where V, E, F, C are the number of vertices, edges, faces and cells, is well satisfied.

Each vertex has 12 neighbours which form an icosahedral shell. Together they form a cluster of 20 tetrahedra all sharing the central vertex.

The dual of the $\{3, 3, 5\}$ polytope is the $\{5, 3, 3\}$ polytope, with 600 vertices, 120 edges, 720 pentagonal faces, and 120 dodecahedral cells.

Note that an alternative description of the $\{3, 3, 5\}$ polytope using the $\{3, 4, 3\}$ polytope (called the 'Gosset' construction), is given in §A6.2.3.

A5.1.2 Shelling and sections

We now proceed to a shell-by-shell description of the $\{3, 3, 5\}$. Indeed, as a polyhedron can be described by gathering vertices on successive coaxial

248

Table A5.1. *Sections of the {3, 3, 5} polytope (with an edge length equal to $2\tau^{-1}$) beginning with a vertex*

Section	x_0	$(x_1, x_2, x_3)^\dagger$	Vertex number	Shape
0	2	$(0, 0, 0)$	1	point
1	τ	$(1, 0, \tau^{-1})$	12	icosahedron
2	1	$(1, 1, 1)$	20	dodecahedron
		$(\tau, \tau^{-1}, 0)$		
3	τ^{-1}	$(\tau, 0, 1)$	12	icosahedron
4	0	$(2, 0, 0)$	30	icosidodecahedron
		$(\tau, 1, \tau^{-1})$		
5	$-\tau^{-1}$	$(\tau, 0, 1)$	12	icosahedron
6	-1	$(1, 1, 1)$	20	dodecahedron
		$(\tau, \tau^{-1}, 0)$		
7	$-\tau$	$(1, 0, \tau^{-1})$	12	icosahedron
8	-2	$(0, 0, 0)$	1	point

$\dagger$Cyclic permutation with all possible changes of signs. $\tau = (1 + \sqrt{5})/2$.

circles, it is possible to describe a polytope as a succession of concentric shells of increasing radius, up to the equatorial sphere, followed by concentric shells of decreasing radius down to the opposite 'south' pole.

Since it is a regular polytope, all sites are equivalent and we can therefore restrict ourselves to one single site which we place at the 'north pole' ($x_0 = R$, $x_1 = x_2 = x_3 = 0$). 'Horizontal' sections are then defined by fixing x_0 or the hyper-polar angle ω (which are related by $x_0 = R\cos\omega$, see appendix A1); indeed fixing ω gives a sphere S^2 embedded in S^3. It is the section of S^3 by a 'horizontal' hyperplane (defined by a constant x_0), similar in lower dimension to the 'parallel' circles on S^2 which are sections of S^2 by horizontal planes.

The 'north pole' is chosen on a vertex of the polytope and so the first shell is a regular icosahedron, and all other shells share the icosahedral symmetry. We get for the following values of ω (see table A5.1 and figure A5.1).

$\omega = \pi/5$: 12 vertices which form a regular icosahedron. These vertices are first neighbours in the polytope.

$\omega = \pi/3$: 20 vertices, a regular dodecahedron. Each vertex is 'above' one triangle of the icosahedron at $\omega = \pi/5$. It is exactly above if one orthogonally maps these two shells on the same 'horizontal' hyperplane.

$\omega = 2\pi/5$: a regular icosahedron whose 12 vertices are second neighbours in the polytope.

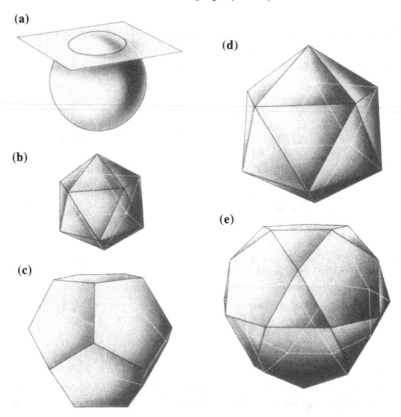

Fig. A5.1. The {3, 3, 5} polytope. Different flat sections in S^3 (with one site on top) give the following successive shells; (a) an icosahedral shell formed by the first 12 neighbours, (b) a dodecahedral shell, (c) a second and larger icosahedral shell, (d) an icosidodecahedral shell on the equatorial sphere. Then other shells are symmetrically disposed in the second 'south' hemi-hypersphere, relative to the equatorial sphere (e).

$\omega = \pi/2$: the 'equatorial' sphere is tiled by 30 vertices which form a regular icosidodecahedron. For larger values of ω, the situation is then symmetrical with respect to the equatorial sphere.

$\omega = 3\pi/5$: an icosahedron.

$\omega = 2\pi/3$: a dodecahedron.

$\omega = 4\pi/5$: an icosahedron.

$\omega = \pi$:   one vertex at the south pole $x_0 = -R$, $x_1 = x_2 = x_3 = 0$.

A5.1.3 Successive shells around a cell centre

Another interesting description consists of fixing a polytope cell centre at the north pole. We get a sequence of polyhedra beginning with two tetrahedra whose respective sizes scale to the golden ratio τ. They are listed in table A5.2.

Table A5.2. *Section of the {3, 3, 5} polytope (edge length $2\tau^{-1}\sqrt{2}$) beginning with a cell*

Section	x_0	(x_1, x_2, x_3)	Vertex number	Shape
0	τ^2	$(\tau^{-1}, \tau-1, \tau^{-1})^\dagger$	4	tetrahedron
1	$\sqrt{5}$	$(-1, 1, 1)$	4	tetrahedron
2	2	$(2, 0, 0)$	6	octahedron
3	τ	(τ, τ, τ^{-2})	12	distorted
4	1	$(\sqrt{5}, 1, 1)$	12	cubo-octahedron
5	τ^{-1}	$(\tau^2, \tau^{-1}, \tau^{-1})$	12	
6	τ^{-2}	(τ, τ, τ)	4	tetrahedron
7	0	$(2, 2, 0)$	12	cubo-octahedron
8	$-\tau^{-2}$	$(-\tau, \tau, \tau)$	4	tetrahedron
—	—	—	—	
14	$-\tau^2$	$(-\tau^{-1}, \tau^{-1}, \tau^{-1})$	4	tetrahedron

$\dagger$Permutation with an even number of sign changes. $\tau = (1 + \sqrt{5})/2$. Distorted cubo-octahedra are such that their square faces are changed into golden rectangles.

A5.1.4 Symmetry properties

The symmetry properties of polytope {3, 3, 5} play an important role in the present context. This question has already been addressed in appendixes A2 and A4. Let us recall that polytope {3, 3, 5} vertices, considered as a set of 120 unit quaternions, form the binary icosahedral group Y'. One necessary condition is that it contains the identity (1, 0, 0, 0) which means that polytope {3, 3, 5} is placed on S^3 such that one vertex is on the 'north' pole (1, 0, 0, 0).

Now the group G' of orientation preserving symmetry operations of polytope {3, 3, 5} is of order 7200. The full symmetry group G contains in addition indirect orthogonal transformations and has order 14 400. This number can be calculated easily using geometric arguments. Indeed the order of the symmetry group of a regular polytope is equal to the number of fundamental regions (or orthoschemes) defined as follows. A given tetrahedral cell of the polytope is decomposed into 24 orthoschemes, which amounts to the required total number of 14 400 (recall that there are 600 tetrahedral cells altogether). The greatest Abelian sub-group of the polytope group contains 30 elements; the orbit can be mapped on a particular 2-plane along a regular 30-gon while, mapped onto a plane orthogonal to the previous one, it gives a 30/11 star polygon (figure A5.2). This 30/11 symmetry is used in chapter 6 in order to compute the spectral properties of the {3, 3, 5} polytope. Its orbit can also be seen as a closed helix, since the 120 polytope vertices can be grouped into four

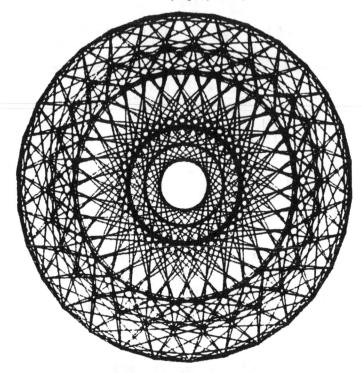

Fig. A5.2. Two-dimensional mapping of the {3, 3, 5} polytope, along the 30/11 symmetry axis.

symmetry related sets of 30 sites, whose local order is a linear arrangement of tetrahedra resembling Coxeter's 'simplicial helix'.

A5.2 Description in terms of toroidal shells

Let us look more deeply into the geometry of the polytope. It is possible to describe the {3, 3, 5} polytope using the so-called spherical torus which is a two-dimensional surface embedded in the spherical space S^3 (appendix A1). Recall that this torus can be built from a square sheet, whose opposite sides are joined together.

Any line parallel to a diagonal of the square corresponds to a great circle of the 3-sphere (a geodesic line). There are two other important lines, the two 'axes' of the torus. For a classical torus one axis is a straight line, the other being a circle. For the spherical torus the two lines are great circles of the 3-sphere.

Let us now proceed to a toroidal decomposition of the {3, 3, 5} polytope. It has two sets of 10 vertices placed on the two opposite axes of the torus foliation. The remaining 100 vertices belong to two sets of 50 vertices, forming triangular tilings on two tori which are placed symmetrically with respect to the spherical torus. We can represent one such torus by a cylinder, the two circular basis having to be identified. The vertices form a column which can be obtained by piling up pentagonal antiprisms with respect to an icosahedral environment for the vertices on the axis (figure A5.3a). Note that, if we decompose further the sets of 50 vertices on each torus into five sets of 10 vertices, we eventually get as a whole 12 sets of 10 vertices belonging to 12 great circles, which is nothing but the discretized Hopf fibration already mentioned in appendix A3 and used in the text to introduce disclinations in the

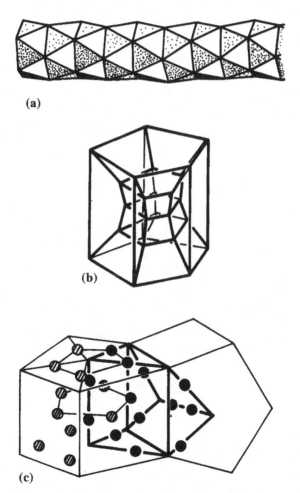

(a)

(b)

(c)

Fig. A5.3. (a) A column obtained by packing pentagonal bi-pyramids. The toroidal layer is obtained if the two ends are identified. (b) Prismatic decomposition of the spherical space with $N = 5$. (c) Position of the $\{3, 3, 5\}$ vertices in this decomposition.

{3, 3, 5} and to calculate its excitation spectrum. The Hopf mapping of this discrete set onto S^2 gives 12 points which form a regular icosahedron on the base.

It is also possible to divide the spherical space into $2N$ prisms having N-sided polygonal bases. Figure A5.3b shows such a representation of this partition with $N = 5$. This representation must be understood as a stereographic map: the outside of the figure is a pentagonal regular prism exactly like other prisms appearing in this figure. To fulfil the partition, each prism is divided into $2N$ triangular prisms as shown in figure A5.3c. Each prismatic unit contains some points of the {3, 3, 5} as shown in the figure.

A6

Frank and Kasper coordination polyhedra

A6.1 Frank and Kasper polyhedra

One can always, in an unambiguous way, divide up a structure, made of points, into (usually irregular) tetrahedra. This is done using first the Voronoi (or Dirichlet) decomposition of space into individual cells which contain the regions of space, closer to a given point than to any other one. In generic cases, the Voronoi cells have three faces sharing a vertex of the cell. Then, connecting the original points of the set whenever their associated Voronoi cells share a face, defines a unique decomposition of the space into tetrahedra. This simplicial decomposition is equivalent, in three dimensions, to a point set triangulation in two dimensions. This procedure also provides the best way to define the coordination number in dense structure: it is the number of faces of the Voronoi cell. In a topological sense the Voronoi cell and the coordination polyhedra are dual. In a tetrahedral division of space, the set of vertices closest to a given site form its first coordination shell, which is a triangulated polyhedron (a deltahedron).

Let us introduce now a standard notation for a site coordination in a tetrahedrally close-packed structure. If the tetrahedra are not too distorted, we can only find situations where either five or six tetrahedra share a given edge. This is the case considered by Frank and Kasper (1958), who then proposed the following notation: a site such that its first neighbour shell is an icosahedron (allowing for small distortions) is called a Z_{12} site. Similarly we define additional sites, denoted Z_{14}, Z_{15}, Z_{16} sites (figure A6.1) according to their number of neighbours. Their corresponding coordination shells are deltahedra with 12 five-fold coordinated and two, three or four (respectively) six-fold coordinated vertices. It can be proven that these additional sites cannot occur isolated in a network, but should form rings or lines (if only Z_{14} sites are invoked) or more generally a subnetwork called the major skeleton or defect network (Sadoc 1983). The edges of these networks are precisely those edges

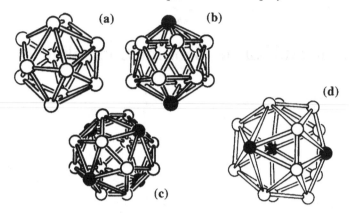

Fig. A6.1. Some Frank–Kasper coordination polyhedra: (a) Z_{12}, (b) Z_{14}, (c) Z_{16}, (d) Z_{15}.

in the simplicial decomposition which share six tetrahedral cells. This topological property of Frank–Kasper structures is related to the non-existence of canonical Z_{13} sites. In a close-packed structure, it is in principle possible to have 13-fold coordinated sites. But this is not compatible then with the above requirement of having only five-fold and six-fold coordinated sites on the coordinated polyhedron, as can be seen in figure A6.2.

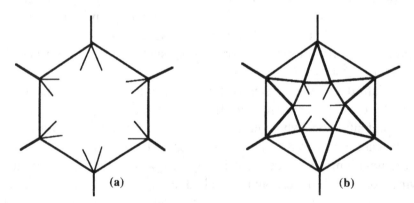

Fig. A6.2. The impossible deltahedron with one six-fold and 12 five-fold vertices. Suppose that this deltahedron exists and map it stereographically onto R^2 with the six-fold vertex as a pole. This vertex is then sent to infinity and (a) displays its six neighbours with the outgoing six edges. As these six vertices are five-fold coordinated, one has two edges per site going inwards to the remaining six vertices. The only solution is to create a new six-fold ring of sites inside (b), which add up to the 13 vertices. It can now be checked that perfect five-fold coordination cannot be achieved for these six new vertices.

A6.2 Positive and negative disclinations

A6.2.1 Other coordination polyhedra with negative disclinations

Let us first briefly recall the distinction between positive and negative disclinations (see §4.1.2). After a Volterra cut, removing matter before gluing creates a positive disclination (a positive concentration of curvature), while adding matter creates a negative disclination.

Let us now describe the above Z_n Frank–Kasper types of sites as disclinated sites. Because we have in mind icosahedral order in tetrahedrally close-packed structures and by reference to the polytope ideal order, the Z_{12} icosahedral site is taken as the reference (undisclinated) state. We have already seen in chapter 4 that a Z_{14} site is obtained by a single Volterra process in a Z_{12} site, the disclination line threading the polyhedron at two opposite points.

We shall also include configurations leading to more distorted tetrahedra, with four tetrahedra around an edge (which then add positive curvature, leading to a positive disclination), and which allow then for 13-fold coordinated sites. The latter can be described topologically by considering the Voronoi cell for a Z_{12} site (a dodecahedron) and adding a new face by dividing a pentagon into two faces. We obtain a new Voronoi cell which has $F = 13$, $E = 33$, $V = 22$. The 13 faces are one square, two hexagons and ten pentagons. Note that this satisfies the Euler relation and equation (4.3.9) in chapter 4.

In order to avoid confusion with the standard Frank–Kasper notation, we propose to call such a configuration Z'_{13}. It is a node threaded by one negative and two positive disclination segments. If we include these new types of lines it is then possible to define other coordinations.

- Z'_{10} if such a line goes through a site (transforming a five-fold axis into a four-fold axis).
- Z'_9 appears when three half negative disclinations cross at the central site. This is in fact a so-called 'Gaskell prism', a configuration which is frequently invoked in amorphous metals. It is formed by a triangular prism with its square faces capped by pyramids (Gaskell 1979).
- Z'_{11} can also be obtained. Starting from a Voronoi cell with z faces, it is possible to add one additional face. One edge and two vertices are added in order to divide a face (figure A6.3). Doing so with a Z'_{10} Voronoi cell (two squares, eight pentagons), we obtain a new cell with three squares, six pentagons and two hexagons. This configuration corresponds to a rather complex node of positive and negative disclinations.

So we can generate a whole family of coordination polyhedra, including the genuine Z_{12} and Z_{14}, Z_{15}, Z_{16} Frank and Kasper polyhedra, but also the following configurations.

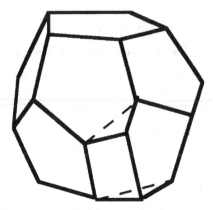

Fig. A6.3. How to increase the coordination by dividing a face of the Voronoi cell. In this case, a pentagonal face of a dodecahedron is divided, adding two sites and three edges. The modified cell has 13 faces, with two hexagons, ten pentagons and one square. There are other solutions: truncation of a vertex (see §7.2), or of an edge.

- Z'_{12}, with a disclination node of two segments of positive and negative disclinations.
- Z'_{14}, which corresponds to the b.c.c. coordination shell: the Voronoi cell is a truncated octahedron with six squares and eight hexagonal faces. There is no pentagon, so it corresponds to a strong distortion as compared to the reference icosahedral coordination.

A6.2.2 Mean coordination number

We have described several structures in §§4.3.3, 5.4 and 7.2 in which atoms have a mean coordination number $\bar{z}$ close to the statistical honeycomb value given by Coxeter, $\bar{z} = 13.39$: recall for instance the value 13.33 found in Laves phases and related hierarchical structures, 13.40 for the β-W structure and even 14 for the b.c.c. structure (which contains Z'_{14} sites).

We have mentioned already that these fluctuations are related to the precise topology of the disclination network. There are indeed two antagonistic effects.

- Large coordination numbers for some sites (Z_{16}, Z_{18} and even more) paradoxically decrease the mean coordination. For instance, compare the Laves phase and the β-W structures (which contain Z_{14} sites).
- Local asymmetries resulting from negative disclinations largely increase the mean coordination. This is observed in disordered structures and also in the b.c.c. structure. In the extreme disordered case of a gas configuration, upon simple division, a value greater than 15 is found for $\bar{z}$ (Matzke 1946).

A6.2.3 An example of an intricate disclination network generating disorder

Let us now look at an example showing how a combination of positive and negative disclination segments can generate disorder. We start with the perfect, curved, close-packed structure, the {3, 3, 5} polytope. The 'pelota' (wool ball) of disclination segments which we are going to introduce does not change the number of vertices, and consequently keeps the total curvature unchanged, even though this new structure will display local curvature fluctuation, responsible for the disorder. We use a special description of the {3, 3, 5} polytope, known as Gosset's construction (Coxeter 1973a, b). Consider a {3, 4, 3} polytope, which is a regular packing of octahedra on S^3. We then generate a new polytope, whose vertices are located at midpoints of the {3, 4, 3} edges. This new structure is a semi-regular polytope, denoted $\{\frac{3}{4}, 3\}$, and is a packing of cubo-octahedra and cubes. The relation between an octahedron and a cubo-octahedron is displayed in figure A6.4a.

Now, consider a distorted form of this new polytope which amounts to displacing consistently the new vertices on the edges of {3, 4, 3}, away from the mid-edges, but in the ratio $1/\tau$. This changes the cubo-octahedra into icosahedra (figure A6.4b), and distorts each cube into five regular tetrahedra (figure A6.5). This new polytope, called the snub-{3, 4, 3}, contains 96 vertices, and is a packing of 120 tetrahedra and 24 icosahedra. Upon adding 24 new vertices in the centres of the icosahedra, a {3, 3, 5} polytope is obtained.

The following description applies to the snub-{3, 4, 3}. From a topological point of view, that is if we only focus on connections between sites, rather than on the distances, we can consider the snub-{3, 4, 3} as the semi-regular $\{\frac{3}{4}, 3\}$ polytope, in which some square diagonals are added as new edges. One rule

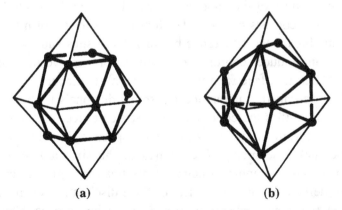

(a)   **(b)**

Fig. A6.4. Relation between an octahedron and (a) the cubo-octahedron or (b) the icosahedron.

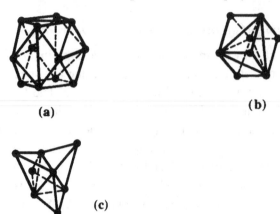

(a)　　　　　　　　　　　　　　　　　　　**(b)**

(c)

Fig. A6.5. (a) A topological icosahedron is obtained by drawing one diagonal on each square of a cubo-octahedron. (b) Same construction on a cube, defined by continuity in the $\{\frac{3}{4}, 3\}$ (a polytope obtained by packing octahedra and cubes). (c) The snub-$\{3, 4, 3\}$ is a packing of icosahedra, connected by the displayed five tetrahedra arrangement, which has been derived from the cube (b) by making all new edges of equal length.

has to be followed: on a cubo-octahedron surface all vertices are five-fold connected. If this is done on every cubo-octahedron, it then extends to the cubes, and by continuity to all the structure, leading to a topological $\{3, 3, 5\}$ (vertices are added at the centre of the topological icosahedra). If all edge lengths are made equal, this topological $\{3, 3, 5\}$ then relaxes into a regular $\{3, 3, 5\}$ polytope.

Now suppose that one mistake is made in the selection of the five tetrahedra inside a cube; this arrangement of tetrahedra is made of a central one capped by four others, and is then completely defined by stating which four vertices, among the eight vertices of the cube, are selected to form this central tetrahedron. The 'mistake' consists of taking the four other sites to form this central tetrahedron. This breaks the coherence between these sites and the remainder of the polytope, and induces coordination changes for some vertices of the six cubo-octahedra sharing a face with the cube. (Recall that cubo-octahedra turn into topological icosahedra upon adding some diagonals.) Flipping some square diagonals modifies the coordination of some vertices on the surface of the cubo-octahedra from five to six or four. By inspection, it is then possible to locate the disclination segments. The centres of the six cubo-octahedra are nodes where two positive and two negative disclination segments cross. The vertices of the 'defective' cube are nodes for three disclination segments of the same sign, and this sign changes alternately on cube vertices. This is one example of what we call a 'pelota' of disclination segments (figure A6.6). The

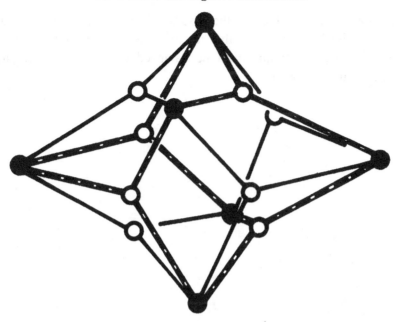

Fig. A6.6. The disclination 'pelota' (wool ball), which appears if the diagonals of a cube (figure A6.5b) are chosen wrongly with respect to the rule defining the snub-$\{3, 4, 3\}$ and the $\{3, 3, 5\}$ polytopes in the Gosset construction.

effect of the mistake on each topological icosahedron can alternatively be described as a T_1 (Weaire and Rivier 1984) transformation leading, on the surface of the icosahedron, to a quadrupole of disclinations (two four-coordinated and two six-coordinated vertices), or in three-dimensional space to a quadrupole of four disclination segments running through the centre of the icosahedron. The 'pelota' of figure A6.6 is mechanically stable, the disclination vectors being in equilibrium at each node.

Whereas vertices at the centre of the icosahedron retain the same coordination, the coordination of the vertices of the cube (which is 12 in the absence of defects – by taking account of the diagonals introduced as edges) is modified by the presence of disclination segments. For the cube vertices which are nodes of three negative disclination segments, the coordination becomes 15 (Z_{15} site). For vertices that are nodes of three positive disclination segments the coordination becomes nine (Z_9' site). This is exactly what is called a 'Gaskell (capped) prism', occurring around the metalloid atom in metal–metalloid amorphous structures. The stability of the prism is due to the topological stability of the 'pelota'. This suggests that we should consider

Gaskell prisms as responsible for the disorder in amorphous metals, rather than being the intrinsic units of the structure. Note that to get a correct disclination density, the prisms must be balanced by negative disclinations (Z_{15} sites inside the 'pelota').

A7

Quasiperiodic tilings: cut and projection

A7.1 Cut and projection algorithm

Let us first briefly recall the 'cut and project' method (Elser 1985, Kalugin *et al.* 1985, Duneau and Katz 1985), which allows us, in its canonical form, to generate quasiperiodic tilings in any dimension. It consists in selecting a set of points and p-dimensional facets inside an n-dimensional periodic structure $(n > p)$, the latter belonging to the 'large space'; the selected objects are then mapped onto p-dimensional Euclidean space, called the 'physical – or parallel – space' E, and form a p-dimensional quasiperiodic tiling. The tiling quasiperiodic character is directly linked to the orientation of E in R^n, the possible presence of an orientational order resulting from the E orientation, but also from the rule for point selection. Take the simplest case of tilings derived from hypercubic lattices Z^n in R^n, equipped with the canonical basis $\{\epsilon_1, \ldots, \epsilon_n\}$. The physical space E, and its complementary space E', also called the 'perpendicular space', are defined by the two complementary projectors π and π', such that

$$E = \pi(R^n) \quad E' = \pi'(R^n) \quad R^n = E + E' \tag{A7.1}$$

Let γ be a sub-set of R^n (generally the hypercube in Z^n); one considers the 'strip', or more properly a hypercylinder,

$$S = E + \gamma = \{x + x'/x \in E \text{ and } x' \in \gamma\} \tag{A7.2}$$

One then defines the 'window' W as $W = \pi'(\gamma)$, the projection of γ onto E'. As W coincides with the intersection of S with E', one can write

$$S = E + W = \{x + x'/x \in E \text{ and } x' \in E'\} \tag{A7.3}$$

The structure X in E is then the projection, onto E, of the set of Z^n points located in S:

$$X = \pi(S \cap Z^n) = \pi[(E + \gamma) \cap Z^n] \qquad (A7.4)$$

Using this algorithm, the best known tilings can be generated: the Fibonacci chain in one dimension ($n = 2$, $p = 1$), the octagonal tiling in two dimensions ($n = 4$, $p = 2$), and finally the icosahedral tiling with rhombohedral cells in three dimensions ($n = 6$, $p = 3$). In these three cases, the window W is the projection of the cube γ^n ($n = 2$, 4, 6), onto E', which gives respectively a segment, a regular octagon and a polyhedron with icosahedral symmetry, the triacontahedron. The first, and most famous tiling, the Penrose pentagonal tiling, can also be generated along the same lines, but in a slightly more complicated way as far as the point selection is concerned, from the Z^5 lattice. The quantity $n - p$ (the dimension of E'), is called the tiling 'codimension'. There are variants of the above method, for instance the 'atomic surfaces' approach, often used for the description of realistic structural models. This method, which is equivalent to the previous one, also makes a link with the approach developed in the 1970s by the Dutch school in the description of incommensurate structures (de Wolf 1974, Janner and Janssen 1977).

These approaches allow us, on the one hand, to generate in an efficient way the quasiperiodic structures in the physical space, but also to calculate easily their structure factors in reciprocal space. Therefore, in about 10 years, the improvement in the quality of the samples, together with very powerful theoretical tools, has permitted us to get a remarkably good knowledge (if one thinks of the initial complexity of the problem) of the atomic positions.

In the following, we shall go back to the generation of quasiperiodic structures, starting from their periodic approximants. We use classical results from number theory, associated with the possibility of generating, for any irrational number, some interesting sequences of rationals which converge towards the irrational. The interest is multifold. First, a prior analytical work allows us to suppress, in the cut and project method, the point selection step. Then, any computer generated structure is necessarily a periodic approximant, since computing (at least in its standard form) does not use irrational numbers, which are approximated by rationals. Why should we let the computer choose this rational? In addition, some large cell crystals, like those studied in chapter 7, can often be analysed as approximants. Finally, when calculating properties, it is often useful to have periodic structural models, even with a very large unit cell, to deal with boundary conditions.

We first discuss codimension 1 structures (Mosseri 1988b), for which the approximant unit cell generation amounts to defining a formal dynamical system, and then show how to generalize these results to systems with larger codimensions, and more precisely to the codimension 2 octagonal tiling (Duneau *et al.* 1989).

A7.2 Codimension 1 approximants

A7.2.1 Projection from two to one dimension

The cut and projection method, although very powerful, nevertheless presents the disadvantage of not providing 'explicit' site coordinates, in the form of one-to-one relations between an enumerated set of indices and the sites coordinates. An example of such explicit coordinates is known for one-dimensional quasicrystals:

$$x_l = l + (\alpha - 1).\text{Int}\left(\frac{l\alpha}{\alpha + 1}\right) \tag{A7.5}$$

where 'Int' is the integer part function and α is an irrational. Let us go back to this one-dimensional example, from a geometrical point of view, and try to generate approximant structures. The points are selected in the Z^2 lattice inside the semi-open strip obtained by translating the unit square along a line E of slope $1/\alpha$. These points are then orthogonally mapped onto E with the projector π_α:

$$\pi_\alpha = \frac{1}{\beta^2}\begin{pmatrix} \alpha^2 & \alpha \\ \alpha & 1 \end{pmatrix} \quad \beta = (1 + \alpha^2)^{1/2} \tag{A7.6}$$

Now, let F_p be a sequence of integers such that $\alpha_p = F_{p+1}/F_p$ approximate better and better the irrational α as p grows, and such that

$$F_{p+1}F_{p-1} - F_p^2 = (-1)^p \tag{A7.7}$$

The approximant structures are generated by translating the unit square along the approximant direction E_p, generated by $\mathbf{u}_p = (F_{p+1}, F_p)$, whose norm is the size of the cell (before projection onto E), and which contains $M_p = F_{p+1} + F_p$ sites. To get all the points in the approximant unit cell, we define a new vector, r_p, which connects the origin to the point in the strip whose distance from E_p is minimal (one chooses for the sake of simplicity E and E_p going through the origin). E_p having a rational slope, there is an infinite number of such points, equivalent by translation. One can choose for example r_p of smallest norm:

$$r_p = (F_p, F_{p-1}) \tag{A7.8}$$

It is then easy to show (Mosseri 1988b) that all sites in the approximant unit cell are obtained by iterating a translation along r_p:

$$x_j^p = j\pi_a(r_p)\mathrm{mod}(\pi_a(u_p)) \quad j = 0, \ldots, M_p - 1 \qquad \text{(A7.9)}$$

where the sites are brought inside the same unit cell, thanks to the 'modulo' operator. Let

$$t_p = \alpha F_p + F_{p-1} \qquad \text{(A7.10)}$$

so that the 'explicit' coordinates of the approximant sites then read

$$x_j^p = \frac{1}{\beta}(jt_p.\mathrm{mod}(t_{p+1})) \quad j = 0, \ldots, M_p - 1 \qquad \text{(A7.11)}$$

One recovers the Fibonacci chain approximants, by taking the Fibonacci series as F_n (which satisfies equation (A7.7)). The unit cell $p = 4$ approximant is represented in figure A7.1.

Notice the unusual site numbering, called the 'conumbering', such that sites are ordered with respect to their distance (in R^2) to the line E_p. The site which is closest to the origin has, as conumber, F_{p+1} for even p and F_p for odd p. It is possible to generalize this construction for any irrational number, even if one does not know a series having the property (A7.7). It suffices to use, for $\mathbf{u}_p$, the sequence of 'best' approximants, and to define r_p with the help of an appropriate Farey sequence. It is possible to return to the usual numbering, and going to the limit $p \to \infty$, to get the standard formulas of the type (A7.5) (Mosseri and Bailly 1989).

Note that the quadrilateral defined by $\mathbf{u}_p$ and r_p has unit area. It is therefore a unit cell of the Z^2 lattice. There is a unimodular matrix with integer entries, having α and $-\alpha^{-1}$ as eigenvalues, which, under repeated action, transforms the unit square into this quadrilateral. This property is at the heart of the method developed below in the octagonal tiling case.

Finally, the knowledge of explicit coordinates allows us easily to get the structure factor. Indeed, the calculations do not require that all sites belong to the same unit cell, which allows us then to disregard the modulo operator. The sites form a limited periodic chain, and one easily gets

$$S(q_m) = \frac{\exp(iAM_p) - 1}{\exp(iA) - 1} \qquad \text{(A7.12)}$$

Fig. A7.1. The unit cell $p = 4$ Fibonacci chain approximant. The site numbering is unusual, and is called the 'conumbering'.

with

$$q_m = 2m\pi \frac{\beta}{t_{p+1}} \quad m \in Z \quad \text{and} \quad A = 2m\pi \frac{t_p}{t_{p+1}}$$

A7.2.2 An example of a tiling obtained by projection from three to two dimensions

The preceding construction is easily extended to any codimension 1 approximant. Indeed, with the orthogonal space E' being unidimensional, the sites can be ordered along this direction and generated by the repeated action of a translation r_p. We shall not enter into the details here, and just show such an example in figure A7.2, with a complete unit cell and the first steps of the discrete dynamical system which generate it.

A7.3 Approximants of the octagonal tiling

Let us now leave the codimension 1 tiling and get interested in the more general cases. It is still possible to get explicit site coordinates, but at the price of a more complex analysis (Mosseri *et al.* 1988). The problem is much easier if the asymptotic quasicrystal presents a scale invariance property, called 'inflation'. Note that all real quasicrystals that have been found in Nature (pentagonal, octagonal, decagonal, dodecagonal and icosahedral) share this property. Let us focus on the octagonal tiling. The matrix M, with integral entries

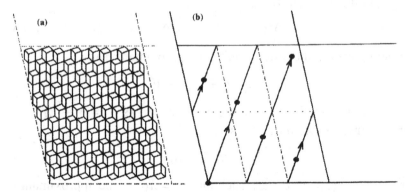

Fig. A7.2. Unit cell of a two-dimensional codimension 1 approximant. Here, $\sqrt{2}$ and the golden ratio have been approximated respectively by $12/7$ and $13/8$, and the structure has been projected onto the approximant plane E_p. (a) A unit cell approximant, containing 343 sites; (b) the first five steps of the dynamical system which generate the sites inside the unit cell.

$$M = \begin{pmatrix} 1 & 0 & 1 & -1 \\ 0 & 1 & 1 & 1 \\ 1 & 1 & 1 & 0 \\ -1 & 1 & 0 & 1 \end{pmatrix} \qquad (A7.13)$$

has eigenvalues (doubly degenerate) $\lambda = 1 + \sqrt{2}$ and $\lambda' = 1 - \sqrt{2}$. M acts on R^4 by letting the octagonal tiling of physical space E be invariant (the eigenspace associated with λ).

This matrix can be written as follows:

$$M = \lambda \pi + \lambda' \pi' \qquad (A7.14)$$

with the two irrational projectors π and π', onto E and E'. In that case, one can also verify that, in addition, M has a unit determinant ± 1. We call this matrix an 'algebraic inflation'. The hyperbolic nature of M allows us to generate a sequence of rational planes E_k which converge toward E. In the present case, we start with the 2-plane E_0 generated by the vectors (ϵ_1, ϵ_2) of the standard orthonormal basis $(\epsilon_1, \epsilon_2, \epsilon_3, \epsilon_4)$. Call L_0 the intersection of Z^4 and E_0. The sequence of rational planes E_k, and the intersection networks L_k are generated by iterating the mapping

$$E_k = M^k(E_0), \quad L_k = Z^n \cap E_k = M^k(L_0) \qquad (A7.15)$$

We introduce the so-called 'Octonacci' sequence $O = 1, 2, 5, 12, 29, \dots$, defined as follows

$$O_{k+1} = 2O_k + O_{k-1}, \quad \text{with} \quad O_1 = 1 \quad \text{and} \quad O_2 = 2 \qquad (A7.16)$$

The successive ratio O_{k+1}/O_k goes towards $\lambda = 1 + \sqrt{2}$. We use the three following parameters α, β, δ:

$$\alpha = (-1)^{k+1} O_k, \quad \delta = (-1)^k O_{k+1}, \quad \beta = \alpha + \delta = (-1)^k (O_k + O_{k-1}) \qquad (A7.17)$$

The number of sites in the order k unit cell approximant reads

$$N_k = 2(|\alpha| + |\beta|)^2 - |\beta|^2 \qquad (A7.18)$$

Let γ be the hypercube in Z^4, and the strip defined as $S_k = \gamma + E_k$. The final approximant structure X_k on E reads

$$X_k = \pi(Z^4 \cap S_k) \qquad (A7.19)$$

Define the vectors (e_1, e_2, e_3, e_4), such that $e_i = \pi(\epsilon_i)$. The approximant lattice reads

$$\lambda^k \pi(L_0) = \lambda^k [e_1, e_2] = [\lambda^k e_1, \lambda^k e_2] \qquad (A7.20)$$

which is simply a dilated square lattice. The next step consists of computing

Table A7.1. *Coordinates for the seven sets of vertices.*
The two first sets I and I' are very similar and only differ
by a simple translation

i	$u_i^1\sqrt{2}$	$u_i^2\sqrt{2}$
I	$2(q+\alpha)$	$2(p+\alpha)$
I'	$2(q+\alpha)+1$	$2(p+\alpha)+1$
II	$p-2q$	p
III	p	$p+2(q+\alpha)$
IV	$-p+2(q+\alpha)+\beta$	$-p+\beta$
V	$p+2\alpha-\beta$	$-p+2(q+\alpha)+\beta$
VI	$p-q+2\alpha$	$p+q$

Table A7.2. *Range for the indices p and q*

i	P_i	Q_i
I	$|\alpha|-1$	$|\alpha|-1$
I'	$|\beta|-1$	$|\alpha|-1$
II	$|\beta|-1$	$|\alpha|-1$
III	$|\beta|-1$	$|\alpha|-1$
IV	$|\beta|-1$	$|\alpha|-1$
V	$|\beta|-1$	$|\alpha|-1$
VI	$p-q+2\alpha$	$p-q+2\alpha$

the explicit coordinates of the N_k sites in a unit cell (see Duneau *et al.* 1989). The net result is a decomposition of the sites into seven families. In each family, the coordinates, in the $[\lambda^k e_1, \lambda^k e_2]$ basis, depend on two integers p and q varying from 0 to P_i and Q_i (depending on the family i and the approximant k). The coordinates (u_i^1, u_i^2) are given in table A7.1 and the corresponding ranges in table A7.2. These coordinates correspond to even values of k. A similar calculation can be done for the odd-k case. Note that, given as such, the sites do not belong to the same unit cell. To get them in a unique cell, it is enough to keep the fractional part of the u_i. The two cases are shown in figure A7.3, with or without the fractional part, for $k = 2$. The case $k = 4$ is presented in figure A7.4.

(a)   (b)

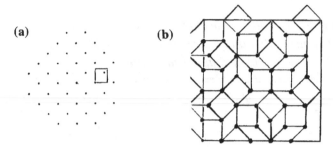

Fig. A7.3. The approximant $k = 2$. (a) An 'extended zone': the sites of a unit cell are in a square lattice limited by an octagon; (b) an approximant unit cell: the sites in (a) have been translated modulo the unit cell.

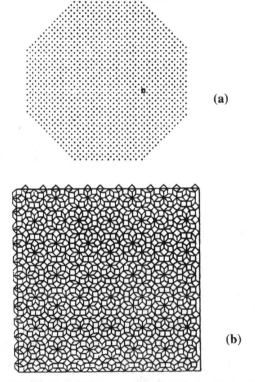

(a)

(b)

Fig. A7.4. The approximant $k = 4$. (a) An 'extended zone': the sites of a unit cell are in a square lattice limited by an octagon; (b) an approximant unit cell: the sites in (a) have been translated modulo the unit cell.

A7.4 An almost octagonal quasiperiodic tiling: the labyrinth

This tiling, composed of three types of tile (and which is a sub-tiling of the standard octagonal tiling) is represented in figure A7.5 (Sire *et al.* 1989). It is remarkable in many respects. First, its vertices are given explicitly and simply as

$$X_{l,m} = l + \alpha(\text{Int}(\alpha(l + m) + 1/2) + \text{Int}(\alpha(l - m) + 1/2))$$

$$Y_{l,m} = m + \alpha(\text{Int}(\alpha(l + m) + 1/2) - \text{Int}(\alpha(l - m) + 1/2))$$

with

$$(l, m) \in Z^2, \quad \alpha = \frac{1}{\sqrt{2}} \tag{A7.21}$$

This tiling also presents inflation properties, and its Fourier spectrum, which has strictly speaking a four-fold symmetry, has an almost eight-fold symmetry,

Fig. A7.5. A piece of the quasiperiodic 'labyrinth' tiling. It contains three types of tile: squares, trapezoids and 'kites'.

which would make it very difficult to distinguish, at an experimental level, from a true octagonal tiling. In the cut and projection framework, it amounts to taking as an acceptance window a square tangent to the standard octagon. Finally, and it is a very attractive point, it is possible to exactly compute its electronic excitation spectrum, with a tight-binding hamiltonian (Sire 1990).

A8

Differential geometry and parallel transport

A8.1 Manifold and tangent space

A manifold, like a sphere or a torus, can be studied from a local point of view by looking at the properties of vectors, or functions, defined in that space. These properties are called intrinsic whenever they are related to quantities that can, in principle, be measured directly inside the manifold, without reference to the embedding space. However, it is often convenient to embed the manifold and use global coordinate systems in the embedding space. The vector normal to a manifold at a given point is an important example of an extrinsic object (which can only be defined with reference to the embedding space).

In physical problems concerned with structural relations, it is quite common to analyse properties using vector fields on the manifold. In order to define vectorial quantities it is necessary to consider the tangent (Euclidean) space of a manifold. A tangent vector at a point $\mathbf{P}$ is an intrinsic local object which can be defined as the tangent vector to a curve $\mathscr{C}$ in the manifold. In terms of differential geometry a vector is a differential operator (derivative) acting on functions. The tangent vector $\mathbf{A}(P)$ at the point $\mathbf{P}(t)$ of a curve parametrized by t is:

$$\mathbf{A}|_P = \frac{\mathrm{d}\mathbf{P}(t)}{\mathrm{d}t} \tag{A8.1}$$

If $\{u^i\}$ is a set of global coordinates then:

$$A^i = \frac{\mathrm{d}u^i}{\mathrm{d}t} \tag{A8.2}$$

Let us consider the derivative of a function f along the curve $\mathscr{C}$. In the following we use the Einstein summation convention on repeated indices: $a^i b_i$ means $\sum_i a^i b_i$. Then:

273

$$\frac{\mathrm{d}f}{\mathrm{d}t}\bigg|_{\mathscr{C}} = \frac{\partial f}{\partial u^i}\frac{\partial u^i}{\partial t} = A^i \frac{\partial}{\partial u^i} f \qquad (A8.3)$$

The vector $\mathbf{A}$ can then be seen as a derivative on the function f in the following way:

$$\mathbf{A}(f) = A^i \frac{\partial}{\partial u^i} f = A^i \mathbf{e}_i(f)$$

where the operator

$$\mathbf{e}_i = \frac{\partial}{\partial u^i} \qquad (A8.4)$$

defines, in the tangent space, the natural frame $\mathbf{e}_i$ with respect to the coordinates. It is also possible to use, in this space, a non-natural frame whose basis vectors are related to the $\mathbf{e}_i$ by a matrix W:

$$\mathbf{e}_\alpha = W^i_\alpha \mathbf{e}_i$$

The vector $\mathbf{A}$ reads:

$$\mathbf{A} = A^\alpha \mathbf{e}_\alpha$$

but, for a generic matrix W, one usually cannot find a coordinate system such that:

$$\mathbf{e}_\alpha = \frac{\partial}{\partial u^\alpha}$$

Let us stress that the tangent vector itself is independent of the choice of coordinates; only its components depend on the choice of the frame, and have a local meaning.

A8.2 Geodesic equation

Geodesic lines play a very important role in a manifold. They are 'the straight lines' of the space, often thought of as minimizing the distance between two close points. In fact, a better definition is to say that a geodesic is a curve whose length has a stationary value with respect to arbitrary small variations of the curve, the end points being held fixed. The length of a curve joining P to Q on the manifold is given by:

$$\mathscr{L} = \int_P^Q (g_{ij}\, \mathrm{d}u^i\, \mathrm{d}u^j)^{1/2} \qquad (A8.5)$$

where the metrics read $g_{ij} = \mathbf{e}_i \mathbf{e}_j$.

A geodesic satisfies the variational condition $\delta \mathscr{L} = 0$, which leads to the geodesic differential equation:

$$\frac{\partial^2 u^i}{\partial t^2} + \gamma^i_{jk} \frac{\partial u^j}{\partial t} \frac{\partial u^k}{\partial t} = 0 \tag{A8.6}$$

The Christoffel symbol γ^i_{jk} carries the information about the space curvature.

The position of indices is important: an upper index indicates that the quantity behaves like the component of a vector (a contravariant object). We can change the position of the indices using the following rules for lifting or lowering

$$\omega_{ij} = g_{jk}\omega^k_i$$

$$\gamma_{ikj} = g_{kl}\gamma^l_{ij} \tag{A8.7}$$

$$\omega^j_i = g^{jk}\omega_{ik}$$

which result from:

$$g_{ik}g^{kj} = \delta^j_i \tag{A8.8}$$

The tensor g^{ij} is defined from the covariant coordinates: it is minor with respect to (i, j) of the matrix g_{ij} divided by the determinant of g.

The Christoffel symbols read:

$$\gamma_{ikj} = \frac{1}{2}\left[\frac{\partial}{\partial u^i}g_{jk} + \frac{\partial}{\partial u^j}g_{ik} - \frac{\partial}{\partial u^k}g_{ij}\right] \tag{A8.9}$$

or, by lifting an index:

$$\gamma^k_{ij} = \frac{1}{2}g^{kl}\left[\frac{\partial}{\partial u^i}g_{jl} + \frac{\partial}{\partial u^j}g_{il} - \frac{\partial}{\partial u^l}g_{ij}\right] \tag{A8.10}$$

A8.3 Parallel transport and curvature

A8.3.1 Parallel transport

Tangent spaces defined at two different points of the manifold are similar linear vector spaces, but they are nevertheless distinct spaces: it is in principle impossible to add two vectors originating at different points. However, we often need to compare vectors at different positions in the manifold. So we need a rule to transport these vectors (figure A8.1).

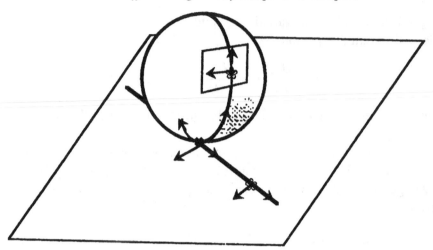

Fig. A8.1. Rolling a sphere on a Euclidean space, a method which allows us to compare two different tangent spaces of S^2: rolling without glide along a line on a Euclidean space. For simplicity we can suppose this line to be a geodesic or a set of successive segments of geodesics.

One particular case of such transport, called the parallel transport of vectors, consists of moving a vector along a geodesic by keeping a constant angle between the vector and the geodesic. But there is still some freedom as the vector can rotate around the geodesic. An approach called 'Schild's ladder', is useful in this context (see Misner *et al.* 1973). In an infinitesimal part of the plane tangent to the geodesic and the given vector, construct a parallelogram, whose sides are the vector and a small displacement along the geodesic. This is extended step by step to the whole geodesic, giving rise to some kind of ladder which allows the transport of the vector.

Consider a vector **a** tangent to a geodesic $u^i(t)$ and a vector field **b**. The variation of **b** along the geodesic is given by an extension of the standard notion of 'derivative', called the covariant derivative:

$$\frac{d\mathbf{b}}{dt} \equiv \nabla_\mathbf{a}\mathbf{b} \tag{A8.11}$$

The parallel transport rule then reads:

$$\nabla_\mathbf{a}\mathbf{b} = 0 \tag{A8.12}$$

which means in the present case that the vector field is parallel along the geodesic line. This definition is not limited to geodesic lines and can be extended to any kind of curve (upon approximating these curves by a sequence of geodesic segments).

In fact, relation (A8.12) is often used to define geodesics. A geodesic is a curve whose tangent vector stays parallel when transported along the curve

$$\nabla_{\mathbf{a}}\mathbf{a} = 0 \qquad (A8.13)$$

A8.3.2 The connection

Consider a vector **a**, defined by its coordinates $a^i = \mathrm{d}u^i/\mathrm{d}t$ in a frame $\mathbf{e}_i = \partial/\partial u^i$. In the calculation of the covariant derivative, one cannot take into account only the coordinate variation, as for the standard derivative in Euclidean space. One must also consider the frame variation, which is then embodied in the so-called 'connection'.

It should be emphasized that a comparison between frames at different points in space is only possible because we have defined a rule for transport of vectors. With the above defined parallel transport, we get a particular, but very important, case of connection, called the Levi–Civita connection. More general connections can result from other transport rules (see for instance the so-called 'double-twist connection', associated with blue phases, chapter 2). For a more general presentation see Dubois-Violette and Pansu (1990) and reference text books like Landau and Lifschitz (1960), Misner *et al.* (1973), or Lovelock and Rund (1975).

Consider a natural frame $\mathscr{R}$ at the point P and locate in this frame another natural frame $\mathscr{R}'$, defined at P', after the latter has been transported parallel at P.

Let us first go back to the choice of local basis vectors at P. If the P coordinates are u^i, the basis vectors of $\mathscr{R}$ are $\mathbf{e}_i = (\partial P/\partial u^i)$. Note that $\mathbf{e}_i$ is a 'true' vector, and not only an operator. Indeed, consider the vector $\mathrm{d}\mathbf{P}$ joining P and P', expressed in the $\mathbf{e}_i$ basis. The differential form $\mathrm{d}\mathbf{P} = \mathrm{d}u^i(\partial P/\partial u^i)$, seen as $\mathrm{d}\mathbf{P} = \mathrm{d}u^i\mathbf{e}_i$, confirms the definition of the basis vectors.

The vectors $\mathbf{e}_i'$ are expressed in the $\mathbf{e}_i$ base, and we can write $\mathrm{d}\mathbf{e}_i = \mathbf{e}_i' - \mathbf{e}_i$:

$$\mathrm{d}\mathbf{e}_i = \omega_i^k \mathbf{e}_k \qquad (A8.14)$$

The change $\mathrm{d}\mathbf{e}_i$ is linearly related to $\mathrm{d}\mathbf{P}$, so the ω_i^k are linear combinations of the $\mathrm{d}u^i$:

$$\omega_i^k = \Gamma_{ij}^k \, \mathrm{d}u^i \qquad (A8.15)$$

The Γ_{ij}^k define the connection.

The covariant derivative can be written with this connection. Using equation (A8.13) to define geodesics, it becomes:

$$\nabla_a a^\alpha = \frac{\partial a^\alpha}{\partial t} + \Gamma^\alpha_{\beta\tau} a^\beta = 0 \qquad (A8.16)$$

The coordinates are expressed in a particular frame chosen in order to have $\mathbf{e}_\tau$ tangent to the geodesic at the point P; the parameter t characterizes the displacement along the geodesics (as in §A8.1, the Greek indices indicate that the frame is not natural). Back in the frame $\mathbf{e}_i$, with $\partial/\partial t = (\partial/\partial u^j)(\partial u^j/\partial t)$, the definition of a geodesic is now:

$$\frac{\partial a^k}{\partial t} + \Gamma^k_{ij} \frac{\partial u^j}{\partial t} a^i = 0$$

which is, using the definition of a^i,

$$\frac{\partial a^k}{\partial t} + \Gamma^k_{ij} \frac{\partial u^j}{\partial t} \frac{\partial u^i}{\partial t} = 0 \qquad (A8.17)$$

By comparison with equation (A8.6), this equation shows that the connection components Γ^k_{ij} are identical to the Christoffel symbols. So we have for the Levi–Civita connection:

$$\Gamma_{ikj} = \frac{1}{2} \left[\frac{\partial}{\partial u^i} g_{jk} + \frac{\partial}{\partial u^j} g_{ik} - \frac{\partial}{\partial u^k} g_{ij} \right]$$

or, upon lifting an index:

$$\Gamma^k_{ij} = \frac{1}{2} g^{kl} \left[\frac{\partial}{\partial u^i} g_{jl} + \frac{\partial}{\partial u^j} g_{il} - \frac{\partial}{\partial u^l} g_{ij} \right]$$

This is exactly the same form as equations (A8.9) and (A8.10).

A8.3.3 Riemann tensor, Ricci tensor and intrinsic curvature

The connection relates the frame at P to the frame at a neighbouring point by

$$\Gamma^j_{pk} = (\nabla_{e_k} \mathbf{e}_p)^j$$

If a vector is transported from one point to another along two different paths, it may happen that its two images are not collinear. The rotation which relates the two orientations indicates the presence of curvature, which can be calculated from the Riemann tensor R^j_{ikl}. The deviation $\delta\mathbf{A}$ of the transported vector along an infinitesimal circuit characterized by the area element $du^k \wedge du^l$ (a 2-form) is:

$$\delta A^i = \sum_j R^i_{jkl} A^j(P) \, du^k \wedge du^l \qquad (A8.18)$$

In this equation we have explicitly indicated the summation, which means that there is no summation on the repeated k and l indices.

From this equation, it is possible to relate the Riemann tensor to the connection. We restrict ourselves to the case of natural frames related to the coordinate system (A8.3). Then,

$$R^j_{ikl} = \frac{\partial \Gamma^j_{ik}}{\partial u^l} - \frac{\partial \Gamma^j_{il}}{\partial u^k} + \Gamma^j_{ml}\Gamma^m_{ik} - \Gamma^j_{mk}\Gamma^m_{il} \tag{A8.19}$$

Note that some authors (like Landau) permutate the k and l indices, which amounts to changing the sign of this tensor.

The Riemann tensor has the following symmetries:

$$R^i_{jkl} = -R^i_{jlk}$$

$$R^i_{jkl} + R^i_{ljk} + R^i_{klj} = 0$$

and, using the metric in order to lower indices:

$$R_{ijkl} = -R_{jikl}$$

$$R_{ijkl} = -R_{ijlk}$$

$$R_{ijkl} = R_{klij}$$

$$R_{ijkl} + R_{iljk} + R_{iklj} = 0$$

The Riemann tensor completely characterizes the local properties of the manifold. It is possible to define simpler tensors, with fewer indices, which therefore bring less information, but nevertheless often enough for a given problem. This is done by what is called a 'contraction', a sum over terms such that the same index appears both in upper and lower positions.

The Ricci tensor is defined as follows:

$$R_{ij} = R^k_{ikj}$$

It is the only possible contraction which does not vanish, due to the symmetries of the Riemann tensor.

It is possible to further contract the Ricci tensor (like the trace of a matrix)

$$\mathscr{R} = g^{ij} R_{ij}$$

This quantity is precisely the space intrinsic scalar curvature.

A8.3.4 Parallel transport around a disclination

Consider a two-dimensional disclination like the one presented in figure A8.2. Suppose that these disclinations are generated in a plane (tiled by squares) on which a field of parallel vectors is defined. A vector transported parallel to the square edges rotates after a closed loop surrounding the disclination. This is a simple example showing the effect of a concentration of curvature on a vector transported by parallel transport. The rotation angle is equal to the angular deficit associated with the disclination.

A8.3.5 Example of Riemann and intrinsic curvature in homogeneous space

Consider, at a point, a local orthogonal frame $(\mathbf{e}_1, \mathbf{e}_2, \mathbf{e}_3)$ and let a vector $\mathbf{A}$ circulate along the curve enclosing the surface element $\mathrm{d}u^1 \, \mathrm{d}u^2$. The vector variation is:

$$\delta A^i = (R^i_{112}A^1 + R^i_{212}A^2 + R^i_{312}A^3) \, \mathrm{d}u^1 \, \mathrm{d}u^2 \tag{A8.20}$$

This variation amounts to a rotation of the vector $\mathbf{A}$, which is characterized

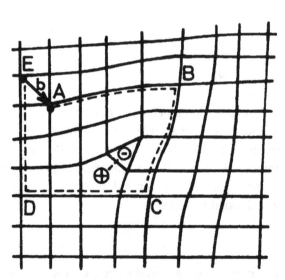

Fig. A8.2. Two disclinations introduced in a square lattice. In one case, a square is changed into a triangle, while in the other it is changed into a pentagon. The Burgers vector is defined along a closed loop surrounding the defects. A vector transported along a loop surrounding only the triangle turns by $\pi/2$, while it turns by $-\pi/2$ around the pentagon. If the path surrounds the two defects, the rotation vanishes, the combined defect being equivalent to a dislocation. This type of defect is related to a characteristic of the space called the space 'torsion', which is not detailed here.

by an axis of rotation orthogonal to the surface element and by an angle ϵ. We can write:

$$\delta A = \left[\begin{pmatrix} \cos\epsilon & -\sin\epsilon & 0 \\ \sin\epsilon & \cos\epsilon & 0 \\ 0 & 0 & 1 \end{pmatrix} - \begin{pmatrix} 1 & 0 & 0 \\ 0 & 1 & 0 \\ 0 & 0 & 1 \end{pmatrix} \right] A \qquad (A8.21)$$

where ϵ is the infinitesimal rotation angle, or the angular deficit if all the curvature of the surface element $du^1 \, du^2$ of area $d\sigma$ is concentrated on a disclination line. We deduce

$$\frac{1}{d\sigma} \begin{pmatrix} \cos\epsilon - 1 & -\sin\epsilon & 0 \\ \sin\epsilon & \cos\epsilon - 1 & 0 \\ 0 & 0 & 0 \end{pmatrix} = - \begin{pmatrix} R^1_{112} & R^1_{212} & R^1_{312} \\ R^2_{112} & R^2_{212} & R^2_{312} \\ R^3_{112} & R^3_{212} & R^3_{312} \end{pmatrix} \qquad (A8.22)$$

which gives some components of the Riemann tensor; other components are obtained by index permutations or using the tensor symmetries.

Consider a three-dimensional space $S^2 \times R$ and three different surfaces through a point in this space. One such surface is a sphere (S^2) and the other two are ordinary cylinders $R \times C_1$ and $R \times C_2$. C_1 and C_2 are two orthogonal great circles on the sphere S^2 with radius r. This space being homogeneous, three such surfaces can be defined at each point. The three vectors e_1, e_2, e_3 are chosen along the three geodesics common to two of the three surfaces (C_1, C_2 and a straight line R). They define a local orthonormal frame. The geodesics on these surfaces are geodesics of the three-dimensional curved space, which allow us to use, for the vector parallel transport, infinitesimal loops lying entirely on these surfaces. See figure A8.3.

First, let us focus on an infinitesimal surface element on S^2, which is a small spherical disc of infinitesimal radius l and area $d\sigma = \pi l^2$. Using the definition of the Gaussian curvature of a surface (see appendix A1, §4) and introducing $\epsilon = \int \int_\mathscr{D} \kappa \, d\sigma$ with $\kappa = 1/r^2$, we find $\epsilon = \pi l^2/r^2$. This is the rotation angle for a vector transported along the loop, with an axis along the R direction. Using equation (A8.22), and keeping only the first order terms, it becomes

$$\begin{pmatrix} R^1_{112} & R^1_{212} & R^1_{312} \\ R^2_{112} & R^2_{212} & R^2_{312} \\ R^3_{112} & R^3_{212} & R^3_{312} \end{pmatrix} \simeq \begin{pmatrix} 0 & 1/r^2 & 0 \\ -1/r^2 & 0 & 0 \\ 0 & 0 & 0 \end{pmatrix} \qquad (A8.23)$$

Now, consider a loop on a cylinder $R \times C$. It is a surface of zero Gaussian curvature, so a vector transported along this loop does not rotate. Corresponding components in the Riemann tensor therefore vanish: $R^i_{j13} = 0$ for all i and j.

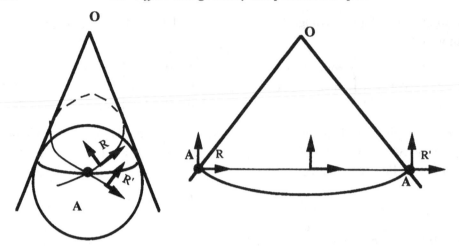

Fig. A8.3. An example of parallel transport and of the non-uniqueness of a local frame in the presence of curvature. The frame at the point A is turned upon a parallel transport along the small circle. In order to define the parallel transport, one considers the cone tangent to the trajectory, which is then flattened. Flat space parallel vectors are then generated, and folded back onto the sphere.

All the components of the Riemann tensor, which are not obtained by symmetry considerations from equation (A8.23), vanish.

The Ricci tensor is obtained by contraction (with $g_{ij} = \delta^j_i$, if one needs R^i_j):

$$\begin{pmatrix} 1/r^2 & 0 & 0 \\ 0 & 1/r^2 & 0 \\ 0 & 0 & 0 \end{pmatrix} \tag{A8.24}$$

The intrinsic curvature is simply the trace of this matrix:

$$\mathscr{R} = 2/r^2$$

which is twice the Gaussian curvature of the 2-sphere $\kappa = 1/r^2$.

Now, consider a sphere S^3 with, at a given point, three orthogonal great spheres S^2 which allow us to define three infinitesimal loops, and lead to the Riemann tensor. We can use the results obtained above on the sphere S^2 (equation (A8.23)), adding permutations, which allow us to write

$$R^1_{212} = R^2_{323} = R^3_{131} = 1/r^2$$

$$R^2_{112} = R^3_{223} = R^1_{331} = -1/r^2$$

$$R^1_{221} = R^2_{332} = R^3_{113} = -1/r^2 \tag{A8.25}$$

$$R^2_{121} = R^3_{232} = R^1_{313} = 1/r^2$$

Then, by contraction, the Ricci tensor reads:

$$\begin{pmatrix} 2/r^2 & 0 & 0 \\ 0 & 2/r^2 & 0 \\ 0 & 0 & 2/r^2 \end{pmatrix} \tag{A8.26}$$

The intrinsic scalar curvature for a spherical space is the trace of the Ricci tensor:

$$\mathscr{R} = 6/r^2$$

A8.3.6 A more intuitive approach to curvature

The Riemann and Ricci tensors, and even the intrinsic scalar curvature, are geometrical objects which are defined analytically; it would clearly be very interesting to have a more intuitive, or synthetic, representation of these objects. We limit our discussion here to a three-dimensional space, but it can be extended to higher dimensions without too many difficulties.

We have seen in appendix A1, §4, following Cartan, that the intrinsic curvature (or the Gaussian curvature) is a quantity which appears when we compare the volume of small elements of the same linear size in Euclidean and curved space. In a space with positive (negative) curvature the volume is smaller (larger) than in a Euclidean space. Intrinsic curvature can also be related to the flux of a field of vectors tangent to all the geodesics issuing from the centre of the volume element through its limiting surface.

The Riemann tensor indicates how a vector is changed when it is transported along a small circuit. Two indices characterize the circuit, another one characterizes the vector, and the fourth one defines the component of the vector variation.

The Ricci tensor, which is an intermediate contraction from the Riemann tensor to the scalar curvature, is not easy to visualize. Let us go back to the homogeneous space presented above.

Consider first a cylinder $S^2 \times R$, with a base S^2 and generatrices which are parallel to R. Then in this space, at a point on the S^2 base, a local frame is defined by tangent vectors to the lines C_1, C_2 and R. C_1 and C_2 are two orthogonal great circles defined on the S^2 base. We consider successively three vector fields of parallel vectors (defined by parallel transport) to a tangent vector to the above three lines at their common point. The first field is parallel to C_1, the second to C_2 and the third to R. We are interested in the variation of the section of the tubes of constant flux in these fields. For instance, the tube of flux for the third field has a constant section: all field lines are straight lines parallel to R and so behave like in a Euclidean space. This can be read directly on the Ricci tensor third diagonal element, which is zero in this example. The first field (and similarly the second one) has field lines which are great circles. Locally, the tube section is a surface orthogonal to C_1, and parametrized by displacements along R and C_2. Take C_1 along a tube axis and a line in the same family on the tube lateral surface, so that the two lines belong to the same S^2 sphere. One can then move along C_2 to relate the two lines in the C_1 family. They are locally parallel near C_2, but being great circles on a 2-sphere they must converge at opposite poles (figure A8.4a). So the section of the flux tube shrinks along the C_2 direction when it is displaced along C_1. In order to analyse better the tube section, let us focus on a line locally parallel to C_1, along a displacement in an orthogonal direction defined by R. The two parallel lines are then two parallel circles on a cylinder $C_1 \times R$, and are equidistant. In that case the tube section does not shrink (figure A8.4b). If the tube section is a small disc, a small displacement along C_1 transforms it into an elliptic disc. This leads to the $1/r^2$ diagonal terms in the Ricci tensor. The other identical term corresponds to the second field parallel to C_2.

Similar arguments can be used in the S^3 example, but in this case a field converges homogeneously in all directions, so the area of a section of a tube flux varies twice as rapidly as in the previous example: all diagonal terms in the Ricci tensor are $2/r^2$.

In conclusion we can say that the Riemann tensor describes the circulation of vectors along closed loops, the Ricci tensor describes the flux of vectors through open $(d-1)$-dimensional surfaces and the intrinsic curvature is related to flux through closed surfaces.

One interesting point in these examples is the property, that is used quite often in this book, that a two-dimensional space can be approximated almost everywhere by a Euclidean space (see figure A8.5), except in small regions of constant intrinsic curvature (locally S^2 or H^2).

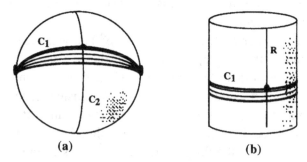

(a)   **(b)**

Fig. A8.4. Schematic drawing showing the behaviour of the flux tubes in two examples of surfaces in the $S^2 \times R$ space. (a) Converging lines, (b) parallel lines.

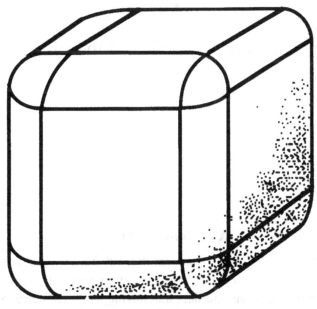

Fig. A8.5. An example of curvature distribution in two dimensions. The edges, which are local discontinuities, are transformed into cylindrical ($S^1 \times R$) regions, with a Euclidean metric: this shows that edges do not carry intrinsic curvature. Then, all the curvature of this surface derived from a cube, is concentrated on the spherical cups close to vertices.

Similarly, a three-dimensional manifold can be approximated almost everywhere by a Euclidean 3-space, except in small regions of constant curvature (locally S^3 or H^3), or close to $S^2 \times R$ or $H^2 \times R$.

If curvature is fully concentrated, it is localized on lines (disclinations) and at nodes where the lines cross. Intrinsic curvature does not distinguish between $S^2 \times R$ (lines) and S^3 or H^3 (nodes); such distinction requires the Ricci tensor.

The Riemann tensor does not bring any new information in three dimensions. In four dimensions, one can similarly locate curvature on regions similar to S^4 (or H^4), $S^3 \times R$ (or $H^3 \times R$), which are natural generalizations of the spaces already described in three dimensions. But one should also consider new types of space, like $S^2 \times R^2$ (or $H^2 \times R^2$). The Riemann tensor allows us now to distinguish between these different cases. In contrast, in two dimensions, the intrinsic curvature (Gauss curvature) fully characterizes the space curvature.

A9

Icosahedral quasicrystals and the $E8$ lattice

A9.1 Introduction

We now consider quasiperiodic structures derived from the eight-dimensional lattice $E8$. Indeed, using the cut and projection method, it is possible to generate a four-dimensional quasicrystal having the symmetry of polytope $\{3, 3, 5\}$ (Elser and Sloane 1987). We present here a modified version of this method. The network is foliated into successive shells surrounding a vertex. These shells belong to S^7 spheres. We then take advantage of the Hopf fibration of S^7, with S^3 fibres, to split the $E8$ sites into symmetrically disposed sets of 24 sites in the S^3 fibres. This method has two advantages: first, in the selection process, it is a full fibre which is either selected or rejected, which simplifies this algorithmic step; more important, this selection process eventually amounts to a simple arithmetic criterion. We then get a shell-by-shell analysis of the four-dimensional structure, which recalls in some respects the 2d–1d algorithm used to generate the Fibonacci chain. The number of points on these shells is exactly given. Note finally that, making sections in this quasicrystal, lower-dimensional quasicrystals are obtained, for instance structures with icosahedral or tetrahedral symmetries in three dimensions. Here we will only summarize these different results, the detailed results being found in the reference (Sadoc and Mosseri 1993).

A9.2 The $E8$ lattice

The $E8$ lattice is known to provide the densest sphere packing in eight dimensions; it belongs to the family of 'laminated' lattices, and is sometimes denoted Λ_8 in this context. The laminated Λ_i form a series starting with the triangular lattice in two dimensions, corresponding to the densest disc packing in the plane. Λ_3 is obtained by suitably packing copies of Λ_2 in the third

286

dimension: one gets the face centred cubic lattice, also known to realize the densest sphere packing in three dimensions. Packing the Λ_3 in the fourth dimension leads to the Λ_4 lattice, also denoted $\{3, 3, 4, 3\}$ as a four-dimensional honeycomb (Coxeter 1973a, b). Upon iterating this construction, one eventually generates the $E8$ lattice.

Consider a (hyper)cubic cell in R^8, with the standard orthonormal basis ϵ_i. Inside this cell, the $E8$ nodes are given by the permutations of

$$(0^8)\left(\frac{1^2}{2}, 0^6\right)\left(\frac{1^4}{2}, 0^4\right)\left(\frac{1^6}{2}, 0^2\right)\left(\frac{1^8}{2}\right) \tag{A9.1}$$

and the preceding set translated by $\left(\frac{1}{4^8}\right)$. Described this way, $E8$ presents an analogy with the f.c.c. lattice and the $\{3, 3, 4, 3\}$ (all the even-dimensional faces are centred), but also with the diamond structure. This cubic cell contains 256 sites.

A9.2.1 The first coordination shell

The shortest distance between two nodes is $\sqrt{2}/2$. Near the origin, one finds sites of the type $(1/2, 1/2, 0^6)$, with all sign combinations, which amounts to 112 sites, and 128 sites of the type $\left(\frac{1^8}{4}\right)$, with the appropriate sign. This first coordination shell is a 240 vertex semi-regular polytope, called the Gosset polytope (Coxeter 1973a, b). Its seven-dimensional facets are, respectively, 17 280 simplices, and 2160 'cross' polytopes. All the lower-dimensional facets are simplices.

A9.3 A discrete Hopf fibration on the Gosset polytope

The 240 vertices of the Gosset polytope belong to a S^7 sphere. It is then interesting to split its 240 vertices into ten equivalent sub-sets, each one belonging to a sphere S^3, and not intersecting the nine others. This is nothing other than a discrete version of the Hopf fibration of S^7 with S^3 fibres and a S^4 base (see appendix A3).

We follow here the presentation given by Manton (1987) of a discrete Hopf fibration.

Consider the elements of the Hurwitz T group (appendix A2), normalized in such a way as to belong to the sphere S^3 of radius $\frac{1}{2}$, and forming a set denoted T_1:

$$T_1 = \left\{ \pm\frac{1}{2}, \pm\frac{\mathbf{i}}{2}, \pm\frac{\mathbf{j}}{2}, \pm\frac{\mathbf{k}}{2}, \frac{1}{4}(\pm 1 + \pm\mathbf{i} + \pm\mathbf{j} + \pm\mathbf{k}) \right\} \tag{A9.2}$$

Let us define then a second set T_2:

$$T_2 = \left\{ \frac{1}{2}(\pm 1 \pm \mathbf{i}), \frac{1}{2}(\pm 1 \pm \mathbf{j}), \frac{1}{2}(\pm 1 \pm \mathbf{k}), \frac{1}{2}(\pm \mathbf{i} \pm \mathbf{j}), \frac{1}{2}(\pm \mathbf{i} \pm \mathbf{k}), \frac{1}{2}(\pm \mathbf{j} \pm \mathbf{k}) \right\}$$

$$(A9.3)$$

T_2 is obtained upon left multiplication of T_1 by the quaternion $1 + \mathbf{i}$. Considered as a set of points on S^3, it forms a polytope $\{3, 4, 3\}$, dual to the one associated with T_1. It is now easy to verify that the 240 vertices of the Gosset polytope can be split into the ten sets:

$$S_1 = (T_2, 0), \quad S_2 = (0, T_2,), \quad S_{3,4} = (T_1, \pm T_1), \quad S_{5,6} = (T_1, \pm \mathbf{i}T_1),$$

$$S_{7,8} = (T_1, \pm \mathbf{j}T_1), \quad S_{9,10} = (T_1, \pm \mathbf{k}T_1)$$

$$(A9.4)$$

This notation should be interpreted as follows. When two numbers appear as indices in S, the first one (or the second, respectively) refers to the plus sign (or minus sign, respectively) in the second term of the pair. One therefore considers pairs of identical quaternions in T_1, the second being multiplied on the left by the specified unit (with the correct sign). For the pair (q_1, q_2), it finally corresponds to a point in R^8.

To each of the ten sets of 24 points S_i corresponds an invariant $Q = q_1 q_2^{-1}$ which characterizes a fibre S^3 of the Hopf fibration. The ten values of Q are $(\infty, 0, \pm 1, \pm \mathbf{i}, \pm \mathbf{j}, \pm \mathbf{k})$ and the corresponding points on the base S^4 are then given by the permutations $(\pm 1, 0, 0, 0, 0)$, which form a 'cross' polytope in R^5. On each fibre, the 24 points form a polytope $\{3, 4, 3\}$. Note that we could continue this process, with the Hopf fibration of S^3, with fibres S^1 and base S^2, and fibrate then each of the $\{3, 4, 3\}$ by six great circles going through four points each (Manton 1987). The six points on the base S^2 form again a cross polytope (here an octahedron), which leads therefore to a particularly interesting and symmetrical pattern for this decomposition of the Gosset polytope.

In addition, each fibre (containing a polytope $\{3, 4, 3\}$) generates a four-dimensional sublattice $\{3, 3, 4, 3\}$ of the $E8$ lattice. There are ten equivalent sublattices through the origin, associated with the ten points on the base S^4, images of the fibres under the Hopf map. Let P be a point on the base S^4, with coordinates

$$\frac{1}{\sqrt{5}}(1, 1, 1, 1, 1).$$

It is equidistant from five of the ten above mentioned points. P defines a four-dimensional space E. The mapping of the Gosset polytope onto E produces two sets of five $\{3, 4, 3\}$, each on a different spherical shell S^3 of E surrounding the origin. One then shows that each set of five $\{3, 4, 3\}$ forms a $\{3, 3, 5\}$ polytope, and one finally gets two concentric $\{3, 3, 5\}$ on E. We recover here a result by Elser and Sloane (1987). E, defined by the point P on S^4, plays here the role of the 'physical' or 'parallel' space of quasicrystals. The 'orthogonal' space E' corresponds to the point P', diametrically opposed to P on the base S^4 of the fibration.

In the following, one must distinguish the Hopf map (H-map) (from S^7 onto the base S^4) and the orthogonal map (O-map), from R^8 onto E (the physical space of the quasicrystal).

A9.4 Shelling the quasicrystal

A9.4.1 Successive concentric shells in E8

The number of vertices in the Nth shell surrounding an $E8$ vertex is given by (Conway and Sloane 1988)

$$V_N = 240 \sum_{d \mid j} d^3 \tag{A9.5}$$

where d denotes all the integers dividing $j = N$.

Table A9.1 displays these values for the first ten shells. Notice that the square of a shell radius equals half its number. Since each shell belongs to a sphere S^7, the set of vertices on one shell can be split into subsets belonging to S^3 fibres, as for the Gosset polytope. Each fibre is H-mapped onto a point on the base S^4. It is then interesting to consider two orientations of the $E8$ lattice (with respect to the Hopf fibrations). The first one, denoted C (for 'crystallographic') coincides with that given above. To get the second one, denoted Q (for 'quasicrystalline'), we make a rotation such that the 4-plane which was H-mapped onto P, is now mapped onto the point $(1, 0, 0, 0, 0)$ of S^4. This rotation can be described with a 2×2 quaternionic matrix $\mathcal{M}$, which acts on pairs of quaternions (points of R^8):

$$\mathcal{M} = \begin{pmatrix} \cos \omega & -\sin \omega \, p' \\ \sin \omega \, \overline{p}' & \cos \omega \end{pmatrix} \tag{A9.6}$$

where

$$\omega = \frac{1}{2} \arccos \frac{1}{\sqrt{5}} \quad \text{and} \quad p' = \frac{1}{2}(1 + i + j + k)$$

Note that ω and p' are related to P in the following way. Let p be the stereographic projection, from the base S^4 onto R^4, of the point P. Then

Table A9.1. *Number of vertices on shells*
surrounding the origin in the E8 lattice. The first
shell is a Gosset polytope in eight dimensions

N	Squared radius r^2	Vertices on E8 shell
1	1/2	240
2	1	2160
3	3/2	6720
4	2	17 520
5	5/2	30 240
6	3	60 480
7	7/2	82 560
8	4	140 400
9	9/2	181 680
10	5	272 160

$p = p' \tan \omega$. $\mathcal{M}$ acts on the pair (q_1, q_2) to give the pair (q'_1, q'_2). In the cut and projection framework, q'_1 is the perpendicular space coordinate in E', and q'_2 the parallel space coordinate in E.

A9.4.2 The base of the fibration in the C orientation

Consider the vertices on the Nth shell, with radius

$$r = \sqrt{\frac{N}{2}},$$

surrounding the origin of $E8$. Note first that, for this shell, the point coordinates on the base (upon H-mapping) read

$$x_i = \frac{v_i}{N} \quad \text{with } v_i \in \mathbf{Z} \tag{A9.7}$$

Indeed, consider the first coordinate x_0 on the base. A point (q_1, q_2) is H-mapped onto S^4

$$x_0 = \cos 2\Omega = \frac{q_2 \bar{q}_2 - q_1 \bar{q}_1}{r^2} = 2 \frac{q_2 \bar{q}_2 - q_1 \bar{q}_1}{N} \tag{A9.8}$$

Since $r^2 = q_1 \bar{q}_1 + q_2 \bar{q}_2$, it is then easy to see, by inspection, that the relation (A9.7) is satisfied when $4q_2 \bar{q}_2$ is an integer. The demonstration, for the other x_i, is done using symmetry considerations. The $E8$ point group coincides with that of the Gosset polytope. We have already seen that the Hopf map of the latter gives a 'cross' polytope on S^4. The Hopf map of any $E8$ shell shares the symmetry of the cross polytope, which allows us in particular to permute the coordinate axes. As the base S^4 has unit radius, it becomes

$$N^2 = v_0^2 + v_1^2 + v_2^2 + v_3^2 + v_4^2 \tag{A9.9}$$

One can show that any combination of v_i which satisfies (A9.9) corresponds to a fibre on the $E8$ Nth shell. This result is not trivial, but results from the solution of a more general problem in number theory (Louck and Metropolis 1981). This allows us to enumerate the points on the base (e.g., the number of fibres S^3) for each shell. Indeed, for each shell N, let us multiply the coordinates on the base by N. The points on the base are then sent onto a shell of radius N in the five-dimensional Z^5 lattice. The number c of points on a Z^5 shell is given by Hardy (1920), and reads, in a simplified form for shells of integral radius,

$$c(N) = \sum_{q|N} \beta(q) \tag{A9.10}$$

where q runs over the divisors of N and

$$\beta(q) = 10q^3 \left(1 - \sum_{odd\, d|q} \frac{1}{d^2} \right) \tag{A9.11}$$

where d are the odd divisors of q, and $d < q$.

A9.4.3 The base of the fibration in the Q orientation

Let the matrix $\mathcal{M}$ act on the $E8$ lattice in order to carry the latter in the above defined Q orientation. The fibration structure is rigidly conserved under this transformation. The base of the fibration, in this orientation, is simply rotated as compared with the C orientation, such that the point

$$\frac{1}{\sqrt{5}}(1, 1, 1, 1, 1)$$

in the base S^4 is brought onto the point $(1, 0, 0, 0, 0)$.

The main interest of this transformation is that, for each $E8$ shell, the points on the base are gathered onto 'horizontal' small spheres S^3 (defined by a constant x_0), enumerated by m, and such that

$$x_0 = \frac{n}{N\sqrt{5}} \tag{A9.12}$$

where $n = 2m$ (N even), or $n = 2m + 1$ (N odd), n taking all integral values between M and $-M$, with

$$M = 2 \left\lfloor \frac{N\sqrt{5}}{2} \right\rfloor (N \text{ even}) \quad \text{or} \quad = 2 \left\lfloor \frac{N\sqrt{5} + 1}{2} \right\rfloor (N \text{ odd}) \qquad \text{(A9.13)}$$

where $\lfloor x \rfloor$ is the integral part of x. One then easily finds that

$$M = 2\lfloor N\tau \rfloor - N \quad \text{with } \tau = \frac{1 + \sqrt{5}}{2} \qquad \text{(A9.14)}$$

Each 'horizontal' set corresponds to the $E8$ vertices which are orthogonally mapped (O-map) onto a unique shell, in both the parallel and perpendicular spaces E and E'. Let us note ρ and ρ', the respective radii of this shell in E and E'; ρ (ρ', respectively) equals the norm of q_2' (q_1', respectively). It is easy to establish relations between ρ, ρ' and the coordinate x_0 on the base S^4 of the fibration;

$$\rho^2 = \frac{r^2(1 + x_0)}{2} \quad \rho'^2 = \frac{r^2(1 - x_0)}{2} \qquad \text{(A9.15)}$$

and, since the number of the $E8$ shell satisfies $N = 2r^2$,

$$\rho^2 = \frac{5N + n\sqrt{5}}{20} \quad \rho'^2 = \frac{5N - n\sqrt{5}}{20} \qquad \text{(A9.16)}$$

where $n = 2m$ or $n = 2m + 1$ according to whether N is even or odd.

These shells, on E and E', are made of more or less regular polytopes, which share the $[3, 3, 5]$ symmetry group. We must then decide which shells should be selected to generate the four-dimensional quasicrystal – this means operating the 'selection step' of the cut and projection algorithm. We are not going to detail this step further (see the above mentioned reference), and rather start to describe the quasicrystal itself. The construction algorithm is schematized in figure A9.1.

A9.5 The 2d–1d aspect of the shell-by-shell construction of the quasicrystal

Two characteristic features of the four-dimensional quasicrystal recall the 2d–1d Fibonacci chain construction.

– First, all shells which contain the same sub-set of points, modulo a scale change, have their radius in one-to-one correspondence with point coordinates on a Fibonacci chain.

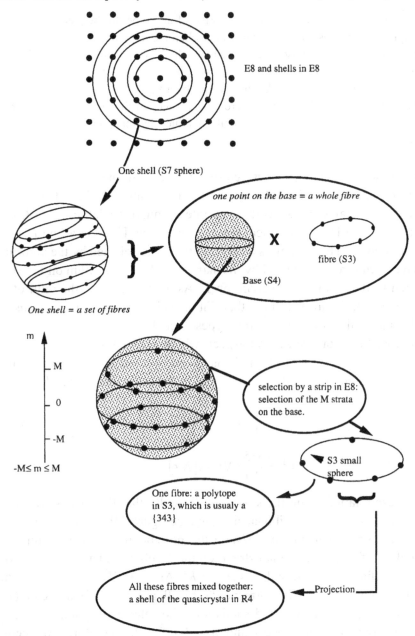

Fig. A9.1. Scheme summarizing the four-dimensional construction method: take an $E8$ shell, considered as a discrete fibration of S^7, select the fibres which map (H-map) onto a stratum M of the base of the fibration, and finally orthogonally map (O-map) the selected sites onto R^4.

- Then, if one takes into account all the shells, of any type, the sequence formed by their square radius also forms a Fibonacci chain. Let us try to understand these properties in the case of those quasicrystal sub-shells which are $\{3, 3, 5\}$ polytopes. The first $E8$ shell (the Gosset polytope) – which is kept as a whole in the selection step – gives two $\{3, 3, 5\}$ in R^4, which differ by a factor τ. Call A_1 (or A_2) the star of 120 vectors in E which point towards the vertices of the smaller (or the larger) $\{3, 3, 5\}$. A_1 generates the whole module obtained by mapping all the $E8$ lattice points.

Consider two vectors, $\mathbf{u}_1$ and $\mathbf{u}_2$, collinear in R^4, and pointing towards two sites, one on each $\{3, 3, 5\}$ polytope. Lifted back into $E8$, these two vectors give two orthogonal vectors, which generate a square sublattice of $E8$, contained in a 2-plane Π. The physical space E intersects Π along a line L which contains u_1 and u_2. The $E8$ points belonging to Π and selected by the algorithm could have been directly selected by the standard 2d–1d algorithm on Π. These points then lead to a Fibonacci chain on L. As a consequence, the radii of the successive $\{3, 3, 5\}$ also follow this Fibonacci sequence. The discussion proceeds in a similar way for the other types of shell.

Let us focus now on the second aspect, which associates a Fibonacci chain with the complete sequence of shells as given by their square radius. These square radii read, in the parallel space E and in the orthogonal space E',

$$\rho^2 = \frac{\sqrt{5}}{10}(N\tau^{-1} + \lfloor N\tau \rfloor) \tag{A9.17}$$

$$\rho'^2 = \frac{\sqrt{5}}{10}(N\tau - \lfloor N\tau \rfloor) \tag{A9.18}$$

From these relations, one finds without difficulty that the successive square radii ρ^2 are ordered along a Fibonacci chain (see figure A9.2). The vertical axis corresponds to the shell number N and the horizontal axis to the number n associated to different 'horizontal strata' on the S^4 fibration base. To each point (N, n) in the figure, corresponds a shell in R^4. As $|n| < N\sqrt{5}$, the full module obtained by mapping all $E8$ corresponds to all points inside a sector ($n = -N\sqrt{5}$, $n = N\sqrt{5}$). The selected shells are associated with points located in between the two lines $n = N\sqrt{5}$ and $n = N\sqrt{5} - 2$. Therefore, we recover here a construction similar to the standard 2d–1d algorithm. But here, all the different types of shell are concerned, and it is their square radius which follows a quasiperiodic order.

It is possible to push the calculations further, and for example to try to determine, in the form of an explicit formula, the number of points on the quasicrystal shells. The exact solution amounts to solving a rather complex

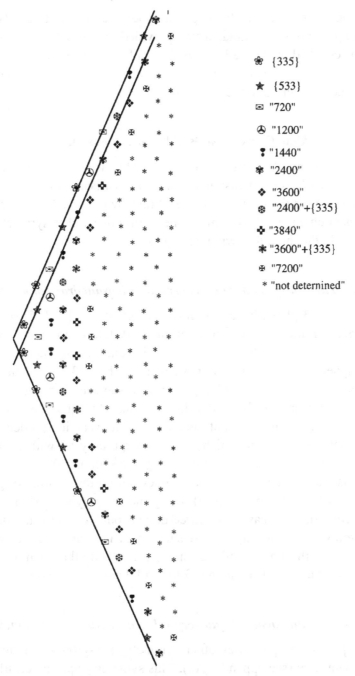

Fig. A9.2. 2d–1d character in the four-dimensional construction of the quasicrystal. Each point represents a shell, a more or less regular polytope. The full set of points inside the sector represents the mapping of all points from $E8$, while the quasicrystal is generated from the shell whose representative points are located in between the two parallel lines, as for the 2d–1d algorithm.

problem, which has to do with deeper number theory questions. We have obtained an approximate formula, which provided the correct value up to the 37th shell (as checked numerically). Moody and Patera (1993) then derived the exact solution, and showed in passing that our approximated formula in fact gives the correct answer up to the 54th shell.

A9.6 Quasicrystals of lower dimension

A rather interesting point in this four-dimensional quasicrystal comes from the fact that interesting lower-dimensional quasicrystals arise, of dimension $d \leqslant 3$, upon cutting by lower-dimensional spaces of high symmetry. A trivial example is the Fibonacci chain, as already mentioned. But other richer symmetries can also appear, as in the three following examples.

A9.6.1 *Three-dimensional quasicrystal of icosahedral symmetry*

One must locate a 3-plane, invariant under the action of a [3, 3, 5] sub-group. The first shell of the four-dimensional quasicrystal, a polytope {3, 3, 5}, belongs to a sphere S^3 which can be cut by a 3-plane, whose intersection with S^3 gives a sphere S^2. An interesting orientation of the 3-plane can then be found by selecting among the polytope vertices, sub-sets forming regular, or semi-regular, polyhedra. We have seen in §A5.1 that the equatorial sphere (S^2) of the polytope – with one vertex at the pole – contains an icosidodecahedron. This sphere defines a 3-plane which contains a quasicrystal with icosahedral symmetry (and even has an icosahedral point group symmetry).

One can also consider a 3-plane which does not go through the origin, and which intersects the polytope {3, 3, 5} along another polyhedron, for example an icosahedron. In that case, the three-dimensional quasicrystal does not possess a vertex at the origin; its first shell is an icosahedron, the second a dodecahedron, the third an icosidodecahedron, ... Note that then, the square radii of the successive shells no longer follow a Fibonacci chain order.

A9.6.2 *Three-dimensional quasicrystals with tetrahedral symmetry*

The {3, 3, 5} polytope possesses other symmetry sub-groups, for instance a tetrahedral symmetry (see appendix A5). This symmetry appears clearly when a cell is centred on a pole. The successive shells, on S^3, are then: a tetrahedron, another tetrahedron inflated by τ, an octahedron, ..., up to a cubo-octahedron on the S^2 equatorial sphere. We have already discussed this sequence in chapter 7, by reference to the Cu–Zn γ phase structure. If one cuts the four-dimensional quasicrystal by a 3-plane which contains this cubo-octahedron, one gets a three-dimensional quasicrystal of tetrahedral symmetry. Note that the first shell is identical to that found in the f.c.c. compact structure.

A9.6.3 Two-dimensional quasicrystals with a 30-fold symmetry

Consider finally a particular case in two dimensions: a quasicrystal presenting a symmetry of order 30. It is well known from classical arguments that such a symmetry can be obtained upon mapping an eight-dimensional lattice (indeed, there are eight integers lower than and prime with 30). The $E8$ point group has a sub-group isomorphic with the [3, 3, 5]. The latter contains a screw symmetry of order 30 (appendix A5). The vertex orbits, mapped onto two orthogonal symmetry planes, lead to either a regular polygon with 30 sides, or a 30/11 star polygon. Upon choosing one of the two planes as 'physical' space, it is then possible to generate a two-dimensional quasicrystal with 30-fold symmetry. However, it is not obtained as a simple cut in the four-dimensional quasicrystal, but requires selection of the points with a certain 'width' around the physical space. More precisely, the acceptance domain in the orthogonal plane is a polygon whose symmetry order is a multiple of 30.

Bibliography

P. M. Ajayan, T. Ichihashi and S. Iijima, *Chem. Phys. Lett.*, **202** (1993) 384.

S. Alexander and J. P. McTague, *Phys. Rev. Lett.*, **41** (1978) 702.

M. Audier and P. Guyot, *Quasicrystalline Materials*, C. Janot and J. M. Dubois, editors, World Scientific (1988) 181.

N. L. Balazs and A. Voros, Chaos on the Pseudosphere, *Physics Reports*, **143(3)** (1986) 110.

R. R. Balmbra, J. S. Clunie and J. F. Goodman, *Nature*, **222** (1969) 1159.

J. Bascle, T. Garel and H. Orland, *J. Physique I France*, **3** (1993) 259.

G. Baskaran, *Phys. Rev.*, **33** (1986) 7594.

S. E. M. Beenker, Univ. of Technology, Eindhoven T.H., Report WSK (1982).

G. Bergman, J. L. T. Waugh and L. Pauling, *Acta Cryst.*, **B 42** (1957) 254.

J. D. Bernal, *Proc. R. Soc. Lond.* **A 280** (1964) 299.

J. L. Black, *Glassy Metals I*, H. Guentherodt and H. Beck, editors, Springer Verlag (1981) 167.

H. W. Blöte and H. J. Hilhorst, *J. Phys. A: Math. Gen.*, **15** (1982) L651.

A. H. Boerdijk, *Philips Res. Rep.*, **7** (1952) 303.

Y. Bouligand, *Geometry in Condensed Matter Physics*, J. F. Sadoc, editor, World Scientific (1990) 191.

M. H. Brodsky and D. P. DiVincenzo, *J. of Non-Crystal. Solids*, **59-60** (1983) 101.

E. Cartan, *Leçons sur la géométrie des espaces de Riemann*, Gauthier-Villars (1963).

J. Charvolin and J.-F. Sadoc, *J. de Physique*, **48** (1987) 1559.

J. Charvolin and J.-F. Sadoc, *J. de Physique*, **49** (1988a) 521.

J. Charvolin and J.-F. Sadoc, *J. Physical Chem.*, **92** (1988b) 5787.

J. Charvolin and J.-F. Sadoc, *Micelles, Membranes, Microemulsions and Monolayers*, W. M. Gelbart, A. Ben-Shaul and D. Roux, editors, Springer-Verlag (1994).

J. Charvolin and A. Tardieu, *Solid State Physics suppl.*, L. Liebert, F. Seitz and D. Turnbull, editors, Academic Press, **14** (1978) 209.

L. Chayes, V. J. Emery, S. A. Kivelsen, Z. Nussinov and G. Tarjus, *Physica*, A **225** (1991), 129.

M. Clerc, A. M. Levelut and J. F. Sadoc, *J. de Physique II France*, **1** (1991) 1263.

G. Coddens and P. Launois, *J. de Phys. I*, **1** (1991) 993.

I. Cohen, A. Ha, X. L. Zhao, M. Lee, T. Fisher, M. J. Strouse and D. Kivelsen, *J. Phys. Chem.* **100** (1991), 8518.

M. H. Cohen and G. S. Grest, *Phys. Rev.*, **B 20** (1979) 1077.

M. Coleman and D. Thomas, *Phys. Stat. Sol.*, **24** (1957) 111.

G. A. N. Connell and R. J. Temkin, *Phys. Rev.*, **B 9** (1974) 5323.

J. H. Conway and N. J. A. Sloane, *Sphere Packings, Lattices and Groups* (Springer Verlag, 1988).

H. S. M. Coxeter, *Illinois J. Math.*, **2** (1958) 746.

H. S. M. Coxeter, *Introduction to Geometry*, Wiley, New York (1969).

H. S. M. Coxeter, *Regular Polytopes*, Dover (1973a).

H. S. M. Coxeter, *Regular Complex Polytopes*, Cambridge Univ. Press (1973b).

H. S. M. Coxeter, *Can. Math. Bull.*, **28** (1985) 385.

H. S. M. Coxeter and W. O. J. Moser, *Generators and Relations for Discrete Groups*, Springer-Berlin (1957).

C. Cros, M. Pouchard and P. Hagenmuller, *C. R. Acad. Sci. Paris*, **260** (1965) 4764.

R. Dandoloff, G., Dohler and H. Bilz, *J. of Non-Crystal. Solids*, **35, 36** (1980) 537.

R. Dandoloff and R. Mosseri, *Europhysics Lett.*, **3** (1987) 1193.

D. W. Davidson, *Water*, F. Franks, editor, Plenum, NY (1973) 115.

P.-G. de Gennes and J. Prost, *The Physics of Liquid Crystals*, Clarendon Press Oxford (1993).

D. P. Deng and M. Widom, *J. Phys. C: Solid St. Phys.*, **20** (1987) L449.

N. Destainville, R. Mosseri and F. Bailly, *Proceeding of the 5th International Conference on Quasicrystals*, C. Janot and R. Mosseri, editors, World Scientific (1995).

R. de Wit, *Int. J. Eng. Sci.*, **19** (1981) 1475.

C. de Witt and B. de Witt, editors, *Relativity, Groups and Topology*, (Les Houches) Gordon and Breach (1963).

P. M. de Wolf, *Acta Crystallogr.*, **A 30** (1974) 777.

D. P. DiVincenzo, *J. de Physique*, **C 47** (1988a) 237.

D. P. DiVincenzo, *Phys. Rev.*, **B 37** (1988b) 1245.

D. P. DiVincenzo and P. J. Steinhardt, editors, *Quasicrystals, the State of the Art*, World Scientific (1991).

J. Dixmier, A. Gheorgiu and M. L. Theye, *J. Phys. C: Solid st. Phys.*, **17** (1984) 2271.

J. Donohue, *The Structure of Elements*, J. Wiley (1974).

E. Dubois-Violette and B. Pansu, *Mol. Cryst. Liq. Cryst.*, **165** (1988) 151.

E. Dubois-Violette and B. Pansu, *Geometry in Condensed Matter Physics*, J. F. Sadoc, editor, World Scientific (1990) 133.

E. Dubois-Violette and B. Pansu, *Mol. Cryst. Liq. Cryst.*, **212** (1992) 225.

E. Dubois-Violette, B. Pansu and P. Pieranski, *Mol. Cryst. Liq. Cryst.*, **192** (1990) 221.

D. M. Duffy and N. Rivier, *J. de Physique Colloque*, **43** (1982) C3-475.

M. Duneau and A. Katz, *Phys. Rev. Lett.*, **54** (1985) 2688.

M. Duneau, R. Mosseri and C. Oguey, *J. Phys. A: Math. Gen.*, **22** (1989) 4549.

M. Duneau and C. Oguey, *Europhys. Lett.*, **13** (1990) 67.

B. Duplantier, *Physica*, **168A** (1990) 179.

P. du Val, *Homographies, Quaternions and Rotations*, Clarendon Press Oxford (1964).

P. Ekwall, in G. H. Brown, editor, Academic Press, *Adv. Liq. Cryst.*, **1** (1975) 1.

S. R. Elliott, *Physics of Amorphous Materials*, Longman, London (1983).

V. Elser, *J. Phys. A: Math. Gen.*, **17** (1984) 1509.

V. Elser, *Phys. Rev.*, **B 32** (1985) 4892.

V. Elser and N. J. A. Sloane, *J. Phys. A: Math. Gen.*, **20** (1987) 6161.

J. Farge, M. F. de Feraudy, B. Raoult and G. Torchet, *J. de Physique Fr.*, **36** (1975) 62.

K. Fontell, *Colloid Polymer Sci.*, **268** (1990) 264.

B. Fourcade, *J. de Physique II Fr.* **2** (1992) 1705.

P. Fowler, *Nature*, **350** (1991) 20.

F. C. Frank and J. S. Kasper, *Acta Crystallogr.*, **11** (1958) 184; **12** (1959) 483.

D. Frenkel, C. L. Henley and E. D. Siggia, *Phys. Rev.*, **B 34** (1986) 3649.

G. Friedel, *Annales de physique,* **18** (1922) 273.

J. Friedel, *Dislocations*, Pergamon Press (1964).

J. Friedel and F. Denoyer, *C. R. Acad. Sci. Paris*, **305** (1987) 171.

C. Furtado and F. Moraes, *Phys. Lett.*, **A 188** (1994) 394.

M. Gardener, *Sci. American*, **236** (1977) 110.

P. Gaskell, *Phil. Mag.*, **B 32** (1975) 211.

P. Gaskell, *J. of Non-Crystal. Sol.*, **32** (1979) 207.

J. P. Gaspard, R. Mosseri and J. F. Sadoc, *The Structure of Non Crystalline Materials*,
    Ph. Gaskell, X. Parker, Y. Davis, editors, Taylor and Francis, London (1982) 550.

J. P. Gaspard, R. Mosseri and J. F. Sadoc, *Phil. Mag.* B **50** (1984) 557.

A. Gheorgiu and M. L. Theye, *Phil. Mag.* B **44** (1981) 285.

B. Grünbaum and G. C. Shepard, *Tilings and Patterns*, Freeman and Co. (1987).

B. I. Halperin and D. R. Nelson, *Phys. Rev. Lett*, **41** (1978) 121.

G. H. Hardy, *Trans. Amer. Math. Soc.*, **21** (1920) 255.

W. F. Harris, *Sci. American*, **237** (1977) 130; *Pour la Science*, **4** (1978), 11.

W. F. Harris, *South African J. of Sc.*, **74** (1978) 332.

D. Hilbert and S. Cohn-Vossen, *Geometry and the Imagination*, Chelsea Pub. Comp.
    New-York (1952).

Lectures on Quasicrystals, F. Hippert and D. Gratias, editors, Éditions de Physique
    Paris (1994).

R. M. Hornreich, M. Kugler and S. Shtrikman, *Phys. Rev. Lett.*, **48** (1982) 1404.

L. Hsu, R. Kusner and J. Sullivan, *Experimental Math.*, **1** (1992) 191.

W. Hume-Rothery, *J. Inst. Metals*, **35** (1926) 295.

A. Janner and T. Janssen, *Phys. Rev.*, **B 15** (1977) 643.

Ch. Janot, *Quasicrystal: a Primer*, Oxford Univ. Press (1992).

C. Janot and M. de Boissieu, *Phys. Rev. Lett.*, **72** (1994) 1664.

M. Jaric, M. Jaric and D. Gratias, editors, *Aperiodicity and Order*, three books,
    Academic Press (1989).

H. Jones, *The Theory of Brillouin Zones and Electronic States in Crystals*, Series in
    Physics, North-Holland Pub. Comp. (1960).

P. Joyes, *Les agrégats inorganiques élémentaires*, Éditions de physique Paris (1990).

P. Jund, D. Caprion, J. F. Sadoc and R. Jullien, *J. Phys.: Condens. Matter*, **9** (1997)
    4051.

P. Kalugin and A. Katz, *Europhys. Lett.*, **21** (1993) 921.

P. A. Kalugin, A. Y. Kitaev and L. C. Levitov, *J. de Physique Lett.*, **46** (1985) L601.

W. Kauzmann, *Chem. Rev.*, **43** (1948) 219.

C. Kittel, *Introduction to Solid State Physics*, John Wiley, New York, London, Sydney
    (1966).

D. Kivelsen, S. A Kivelsen, X. L. Zhao, Z. Nussinov and G. Tarjus, *Physica*, A **219**
    (1995), 15.

F. Klein, *Le programme d'Erlangen*, Gauthier-Villard Paris (1974).

J. Kepler, *Harmonice Mundi* (1619).

M. Kléman, *Points, Lignes et Parois*, Éditions de Physique (1977); *Points, Lines and
    Walls*, J. Wiley (1983).

M. Kléman, *J. de Physique*, **46** (1985) 1193.

M. Kléman, *J. de Physique Lett.*, **46** (1985) L-723.

M. Kléman, *Advances in Phys.*, **38** (1989) 605.

M. Kléman, *Rep. Prog. Phys.*, **52** (1989) 555.

M. Kléman and P. Donnadieu, *Phil. Mag.*, **B52** (1985) 121.

M. Kléman and J. F. Sadoc, *J. de Physique*, **40** (1979) L569.

P. Kramer, *Acta Crystallogr.*, A **38** (1982) 257.

P. Kramer and Z. Papadopoulos, *Can. J. Phys.*, **72** (1994) 408.

E. Kröner, *Physics of Defects*, R. Balian, M. Kleman and J. P. Poirier, editors, North Holland, Amsterdam (1981) 215.

H. Kroto, H. W. Heath, J. R. O'Brien, R. F. Curl and R. E. Smalley, *Nature*, **318** (1985) 162.

R. Kusner, *Pacific Jour. of Math.*, **138** (1989) 317.

R. Kusner, *Proc. R. Soc. London*, **A439** (1992) 683.

B. Laird and H. Schober, *Phys. Rev. Lett.*, **66** (1991) 636.

L. Landau and E. Lifschitz, *Théorie du Champ*, éditions de Moscou (1960).

J. L. Lauriat, *J. of Non-Crystal. Solids*, **55** (1983) 77.

A. Lautmann, *Essai sur l'unité des mathématiques*, collection 10/18 (1977).

T. Lenosky, X. Gonze, M. Teter and V. Elser, *Nature*, **355** (1992) 333.

D. Levine and P. J. Steinhart, *Phys. Rev. Lett.* **53** (1984), 2477.

W. Li, H. Park and M. Widom, *J. Stat. Phys.*, **66** (1992) 1.

D. Lovelock and H. Rund, *Tensors, Differential Forms and Variational Principles*, Wiley and Sons (1975).

J. D. Louck and N. Metropolis, *Adv. Appl. Math*, **2** (1981) 138.

V. Luzzati, *Biological membranes* 1, D. Chapman, editor, Academic Press (1968) 71.

P. A. MacMahon, *Combinatory Analysis*, Cambridge Univ. Press (1916).

A. L. Mackay, *Acta Crystallogr.*, **15** (1962) 916.

A. L. Mackay, *Physica*, **114A** (1982) 609.

A. L. Mackay and H. Terrones, *Nature*, **352** (1991) 762.

W. Magnus, *Non Euclidean Tesselations and their Groups*, Academic Press (1974).

A. Major, N. Taylor, A. Lawrence, N. Rivier and J. F. Sadoc, *Phil. Mag. B*, **55** (1987) 507.

N. S. Manton, *Comm. Math. Phys.*, **113** (1987) 341.

P. Mariani, V. Luzzati and H. Delcroix, *J. Mòl. Biol.*, **204** (1988) 165.

E. B. Matzke, *Am. J. Botany*, **33** (1946) 58.

S. Meiboom, M. Sammon and D. W. Berreman, *Phys. Rev.*, **A 28** (1983) 3553.

D. Mercier and J. C. Levy, *Phys. Rev.*, **B 27** (1983) 1292.

N. D. Mermin, *Rev. Mod. Phys.*, **51** (1979) 591.

X. Michalet and D. Bensimon, *Phys. Rev. Lett.*, **72** (1994) 168.

X. Michalet, F. Jülicher, U. Seifert and D. Bensimon, *La Recherche*, **269** (1994) 631.

J. W. Mintmire, B. I. Dunlap and C. T. White, *Phys. Rev. Lett.*, **68** (1994) 1012.

C. W. Misner, K. S. Thorne and J. A. Wheeler, *Gravitation*, Freeman San Francisco (1973).

R. V. Moody and J. Patera, *J. Phys. A: Math. Gen.*, **26** (1993) 2829.

R. Mosseri, Thèse d'Etat, Université Paris-Sud.

R. Mosseri, *Geometry of Disordered Systems*, in *From Crystal to Amorphous*, C. Godrèche, editor, Ed. de Physique (1988a) 1.

R. Mosseri, *Universalities in condensed matter* (Les Houches), R. Jullien, L. Peliti, R. Rammal and N. Boccara, editors, Springer (1988b) 9.

R. Mosseri, *J. of Non-Crystal. Solids*, **153–154** (1993) 658.

R. Mosseri, *Lectures on Quasicrystals*, F. Hippert and D. Gratias, editors, Editions de Physique (1994).

R. Mosseri and F. Bailly, in *Number Theory and Physics* (Les Houches), Springer (1989).

R. Mosseri and F. Bailly, *Int. J. of Mod. Phys.*, **B 7** (1993) 1427.

R. Mosseri, F. Bailly and C. Sire, *J. of Non-Crystal. Solids*, **153–154** (1993) 201.

R. Mosseri and J. P. Gaspard, *J. of Non-Crystal. Solids*, **97–98** (1987) 415.

R. Mosseri, D. P. DiVincenzo, M. H. Brodsky and J. F. Sadoc, *Phys. Rev.*, **B 32** (1985) 3974.

R. Mosseri, C. Oguey and M. Duneau, *ILL-CODEST Workshop*, C. Janot and J. M. Dubois, editors, World Scientific (1988), 224.

R. Mosseri and J. F. Sadoc, *Structure of Non Crystalline Materials 1982*, P. H. Gaskell, J. H. Parker and E. A. Davis, editors, Taylor and Francis (1982) 137.

R. Mosseri and J. F. Sadoc, *J. de Physique Lett.*, **45** (1984) L827.

R. Mosseri and J. F. Sadoc, *Z. Phys. D*, **12** (1989) 89.

R. Mosseri and J. F. Sadoc, *Geometry in Condensed Matter Physics*, J. F. Sadoc, editor, World Scientific (1990) 235.

R. Mosseri, J. F. Sadoc and J. Charvolin, *The Structure and Conformation of Amphiphilic Membranes*, R. Lipowsky, D. Richter, K. Kremer, editors, Springer Proceedings in Physics, **66** (1992).

R. Mosseri and J. F. Sadoc, *Proceeding of the 5th International Conference on Quasicrystals*, C. Janot and R. Mosseri, editors, World Scientific (1995) 747.

M. Mutz and D. Bensimon, *Phys. Rev.*, **A 43** (1991) 4525.

D. R. Nelson, *The Taniguchi Symposium on Nature of Topological Disorder*, F. Yonezawa and T. Ninomiya, editors, Springer Proceedings in Physics (1982) 1.

D. R. Nelson, *Phys. Rev.*, **B 28** (1983) 5515.

D. R. Nelson and B. I. Halperin, *Phys. Rev.*, **B 19** (1979) 1181.

D. R. Nelson, M. Rubinstein and F. Spaepen, *Phil. Mag.*, **A 46** (1982) 105.

D. R. Nelson and M. Widom, *Nucl. Phys.*, **B 240** (1984) 113.

S. Nicolis, R. Mosseri and J. F. Sadoc, *Europhys. Lett.*, **1** (1986) 571.

S. Nicolis, R. Mosseri and J. F. Sadoc, *J. de Physique*, **49** (1988) 599–604.

J. A. Oberteuffer and J. A. Ibers, *Acta Crystal.*, **26** (1970) 1499.

Ou-Yang and Zhong-Can, *Phys. Rev.*, **A 41** (1990) 4517.

V. Paillard, P. Melinon, V. Dupuis, A. Perez, J. P. Perez, G. Guiraud, J. Formazero and G. Panzer, *Phys. Rev.*, **B 49** (1994) 11 433.

B. Pansu, E. Dubois-Violette and R. Dandoloff, *J. de Physique*, **48** (1987) 1559.

L. Pauling, *Nature of the Chemical Bond*, Cornell Univ. Press (1960).

L. Pauling, *Nature*, **317** (1985) 512.

U. Pinkall and I. Sterling, *Math. Intelligencer*, **9** (1987) 38.

H. Poincaré, *C. R. Acad. Sci. Paris*, **117** (1893) 144.

D. Polk, *J. of Non-Crystal. Solids*, **5** (1971) 365.

T. Regge, *Il Nuovo Cimento*, **29** (1961) 558.

P. Reichert and R. Schilling, *Phys. Rev.*, **B 30** (1984) 917.

B. Riemann, *Gesammelte Mathematische Werke*, (1892) 88.

N. Rivier, *Phil. Mag. B*, **40** (1979) 859.

N. Rivier, *J. de Physique Colloque*, **C9–43** (1982a) 91.

N. Rivier, *Phil. Mag.*, **A 45** (1982b) 1081.

N. Rivier, *the Taniguchi Symposium on Nature of Topological Disorder*, F. Yonezawa and T. Ninomiya, editors, Springer Verlag (1983) 14.

N. Rivier, *Adv. Phys.*, **36** (1987) 95.

N. Rivier, *Geometry in Condensed Matter Physics*, J.-F. Sadoc, editor, World Scientific (1990) 1.

N. Rivier, Galeener symposium, *J. of Non-Crystal. Solids*, **182** (1995) 162.

N. Rivier and D. M. Duffy, *J. de Physique*, **43** (1982) 293.

N. Rivier and J. F. Sadoc, *Europhys. Lett.*, **7** (1988) 526.

G. de Robinson, *Proc. Camb. Phil. Soc.*, **27** (1931) 1931.

M. Roček and R. M. William, *Class. Quantum Grav.*, **2** (1985) 701.

D. Romeu, *Int. Jour. Mod. Phys.*, **B2** (1988) 267 and 77.

Sachdev and D. Nelson, *Phys. Rev. Lett.*, **53** (1984) 1947.

Sachdev and D. Nelson, *Phys. Rev.*, **B 32** (1985) 1480.

J. F. Sadoc, *J. of Non-Crystal. Solids*, **44** (1981a) 1.

J. F. Sadoc, *C.R. Acad. Sc. Paris*, série II **292** (1981b) 435.

J. F. Sadoc, *J. de Physique Lett.*, **44** (1983) L-707.

J. F. Sadoc, *J. de Physique Colloque*, **C9−46** (1985) 79.

J. F. Sadoc and J. Charvolin, *J. de Physique*, **47** (1986) 683.

J. F. Sadoc, J. Dixmier and A. Guinier, *Journ. of Non-Crystal. Solids*, **12** (1973) 48.

J. F. Sadoc and R. Mosseri, *Phil. Mag.*, **45** (1982a) 467.

J. F. Sadoc and R. Mosseri, *J. de Physique Colloque*, **C9−43** (1982b) 97.

J. F. Sadoc and R. Mosseri, *J. de Physique*, **45** (1984) 1025.

J. F. Sadoc and R. Mosseri, *J. de Physique*, **46** (1985a) 1809.

J. F. Sadoc and R. Mosseri, *J. de Physique Colloque*, **C8−46** (1985b) 421.

J. F. Sadoc and R. Mosseri, *J. Phys. A: Math. Gen.*, **26** (1993) 1789.

J. F. Sadoc and N. Rivier, *Phil. Mag.*, **B 55** (1987) 537.

S. Samson, *The Structure of Intermetallic Compounds.* In *Structural Chemistry and Molecular Biology*, A. Rich and N. Davidson, editors, Freeman San Franscisco (1968) 687.

C. B. Schoemaker and D. P. Schoemaker, *Aperiodicity and Order, Tome III: Extended Icosahedral Structures*, M. Jaric and D. Gratias, editors, Academic Press (1989).

P. Scott, *Bull. London Math. Soc.*, **15** (1983) 401.

J. V. Selinger and D. Nelson, *Phase Transitions in Condensed Matter Systems − Experiments and Theory*, G. S. Cargill, F. Spaepen and K. N. Tu, editors, Pittsburgh Materials Research Society (1987) 185.

H. Seifert, *Acta Math.*, **60** (1933) 147.

H. Seifert and W. Threfall, *A Textbook of Topology*, Academic Press (1980).

U. Seifert, *Phys. Rev. Lett.*, **66** (1991) 2404.

U. Seifert, *Phys. Rev. Lett.*, **70** (1993) 1335.

J. P. Sethna, *Phys. Rev. Lett.*, **51** (1983) 2198; *Europhys. Lett.*, **26** (1994) 51.

J. P. Sethna, D. C. Wright and N. D. Mermin, *Phys. Rev. Lett.*, **51** (1983) 467.

D. Shechtman, I. Blech, D. Gratias and J. W. Cahn, *Phys. Rev. Lett.*, **53** (1984) 1951.

J. D. Shore, M. Holzer and J. P. Setha, *Phys. Rev.*, **B 46** (1992), 11 376.

A. K. Sinha, Prog. in mat. sci., Pergamon New-York, **15** part 2 (1972) 79.

C. Sire, thèse, novembre 1990, université Paris 6.

C. Sire and R. Mosseri, Compactness in icosahedral clusters, *Proceedings of the 6th International Conference on Quasicrystals*, Tokyo, May 1997, S. Takeuchi and T. Fujiwara, editors, World Scientific, (1998) 108.

C. Sire and M. Bellissard, *Europhys. Lett.*, **11** (1990) 439.

C. Sire, R. Mosseri and J. F. Sadoc, *J. de Physique*, **50** (1989) 3463.

F. Spaepen and R. B. Meyer, *Scripta Met.*, **10** (1976) 257.

P. J. Steinhart, D. R. Nelson and M. Ronchetti, *Phys. Rev.*, **B 28** (1983) 784.

J. P. Straley, *Mat. Sciences Forum*, **4** (1985) 93.

J. P. Straley, *Phys. Rev.*, **B 34** (1986) 405.

G. Tarjus, D. Kivelsen and S. A. Kivelsen, in *Experimental and Theoretical Approaches to Supercooled Liquids: Advances and Novel Applications*, J. Fourkas, editor, ACS Books (1998), in the press.

W. P. Thurston, *Bull. Amer. Math. Soc.*, **6** (1982) 357.

L. Tilton, *J. Res. Nat. Bur. Standard*, **59** (1957) 139.

G. Toulouse, *Comm. Phys.*, **2** (1977) 115.

G. Toulouse, *Modern Trends in the Theory of Condensed Matter*, A. Pekalski and J. Prystawa, editors, Springer (1980) 195.

G. Toulouse and M. Kléman, *J. Phys. Lett.*, **37** (1976) L149.

H. R. Trebin, *Adv. Phys.*, **31** (1982) 195.

G. Venkataramann and D. Sahoo, *Contemp Phys.*, **26** (1985) 579; *Contemp Phys.*, **27** (1986) 3.

G. Venkataramann, D. Sahoo and V. Balakrishnan, *Beyond the Crystal*, Sol. Stat. Sciences, **84** (1989) Springer.

L. J. Vieland and A. W. Wicklund, *Phys. Lett.*, **49A** (1974) 407.

G. E. Volovik and V. P. Mineev, *Sov. Phys. JETP*, **45** (1977) 1186.

Z. H. Wang and K. M. Kuo, *Acta Crystallogr.*, **A 64** (1988) 857.

G. H. Wannier, *Phys. Rev.*, **79** (1950) 357.

N. P. Warner, *Proc R. Soc. Lond.*, **A 383** (1982) 379.

D. Weaire and N. Rivier, *Contemp. Phys.*, **25** (1984) 59.

I. A. Weerasekera, S. Ismat Shah, D. V. Baxter and K. M. Unruh, *Appl. Phys. Lett.*, **64** (1994) 3231.

A. F. Wells, *Structural Inorganic Chemistry*, Clarendon Press, Oxford (1962).

A. F. Wells, *Three-Dimensional Nets and Polyhedra*, Wiley (1977).

M. Widom, *Phys. Rev.*, **B 34** (1986) 756.

M. Widom, *Aperiodicity and Order, Tome I: Introduction to Quasicrystals*, M. Jaric, editor, Academic Press (1988) 59.

M. Widom, N. Destainville, R. Mosseri and F. Bailly, *Proceedings of the 6th International Conference on Quasicrystals*, Tokyo, May 1997, S. Takeuchi and T. Fujiwara, editors, World Scientific (1998).

R. William, *The Geometrical Foundation of Natural Structure*, Dover (1979).

T. J. Willmore, *Total Curvature in Riemannian Geometry*, Ellis Horwood Ltd. Chichester (1982).

R. R. Winters and W. S. Hammack, *Science*, **260** (1993) 202.

F. Wooten, Galeener symposium, *J. of Non-Crystal. Solids*, **182** (1995) 154.

N. H. Zachariasen, *J. Am. Chem. Soc.*, **54** (1932) 3841.

Index